Special Publication No. 73

Aluminium in Food and the Environment

The Proceedings of a Symposium organised by the Environment and Food Chemistry Groups of the Industrial Division of the Royal Society of Chemistry

London, 17th May 1988

Edited by
R. Massey
Ministry of Agriculture, Fisheries and Food, Norwich

D. Taylor
Imperial Chemical Industries PLC, Brixham

British Library Cataloguing in Publication Data

Aluminium in food and the environment: the proceedings of a symposium organised by the Environment and Food Chemistry Groups of the Royal Society of Chemistry, London, 17th May 1988.—(Special publication, no. 73)
1. Food. Contamination by aluminium 2. Environment. Pollution by aluminium
I. Massey, R. II. Taylor, D. (David) III. Royal Society of Chemistry. *Environmental Group* IV. Royal Society of Chemistry. *Food Chemistry Group* V. Series
363.1′92

ISBN 0-85186-846-0

Published by The Royal Society of Chemistry
Thomas Graham House, The Science Park, Cambridge CB4 4WF

Printed and bound in Great Britain by
Whitstable Litho Printers Ltd., Whitstable, Kent

Preface

This volume contains the proceedings of a one day seminar held at the Scientific Societies Lecture Theatre in London on 17 May 1988.

This meeting was organised jointly by the Environment and the Food Chemistry Groups of the Industrial Division of the Royal Society of Chemistry, and is the first of a series of meetings being arranged to discuss the chemistry of the interrelationships between the environment and the foods we eat.

We should like to thank the authors who have provided papers for inclusion in this volume and to all those who contributed to the success of the original meeting, in particular to the other members of the joint organising committee, C S Creaser, D Henshall and R Purchase.

Dr R Massey
Food Chemistry Group

Dr D Taylor
Environment Group

Contents

Aluminium in Food and the Environment — 1
J.R. Duffield and D.R. Williams

Aluminium Toxicity in Individuals with Chronic Renal Disease — 6
W.K. Stewart

Aluminium and the Pathogenesis of Neurodegenerative Disorders — 20
J.A. Edwardson, A.E. Oakley, R.G.L. Pullen, F.K. McArthur, C.M. Morris, G.A. Taylor, and J.M. Candy

An Epidemiological Approach to Aluminium and Alzheimer's Disease — 37
C.N. Martyn

The Chemistry of Aluminium and Silicon within the Biological Environment — 40
J.D. Birchall and J.S. Chappell

The Determination of Aluminium in Foods and Biological Materials — 52
H.T. Delves, B. Suchak, and C.S. Fellows

Aluminium in Foods and the Diet — 68
J.C. Sherlock

Aluminium in Infant Formulae and Tea and Leaching during Cooking — 77
M.J. Baxter, J.A. Burrell, H.M. Crews, and R.C. Massey

The Use of Aluminium - Especially as Packaging Material - in the Food Industry — 88
H. Severus

Subject Index — 103

Aluminium in Food and the Environment

J.R. Duffield and D.R. Williams

DEPARTMENT OF APPLIED CHEMISTRY, UWIST, PO BOX 13, CARDIFF CF1 3XF, UK

1 INTRODUCTION

Mankind evolved utilising primitive biochemicals and trace metals on the Earth's surface which, in general, were those elements most readily and easily available. This was a protracted process which started more than 4.5 thousand million years ago. However, even though aluminium is the third most abundant element in the Earth's crust, by reason of its chemical nature it has been effectively excluded from normal biochemical and metabolic processes. This is due, in large part, to the low solubility of aluminium silicates, phosphates and oxides rendering the aluminium chemically unavailable. Consideration of these facts can be used to explain the following observations.

- There is no active or specific pathway for the uptake and retention of aluminium by man.

- The two most abundant elements in the lithosphere, before aluminium, are oxygen and silicon. It is, thus, not surprising that aluminium is usually found bonded to silicates in nature. This will also detract from the chemical and biological availability of the element.

- Early medical practice utilised materials taken directly from nature with little, or no, processing and, therefore, many antacids and other medicaments, such as phosphate absorption excluders, are based upon aluminium.

- Given the above, essentiality of aluminium as a dietary factor is highly unlikely. Direct evidence, however, is difficult to obtain due to the manifold difficulties of producing aluminium free diets for experimentation.

2 ALUMINIUM IN THE ENVIRONMENT

Not only is aluminium ubiquitous, and present in large amounts in our environment, but also the wide range of engineering uses found for aluminium has resulted in a world-wide consumption of more than 22 million tonnes in 1986, a figure that is rapidly rising. Without aluminium, our buildings would not function, our kitchen equipment would be heavy and cumbersome, and modern aircraft would have difficulty in getting off the ground. The built-in obsolescence associated with such consumer goods means that, over and above the already aluminium-rich environment in which we live, the disposal of aluminium commodities can further add to the apparent burden for *homo sapiens* (of the 22 million tonnes mentioned above, 16 million are raw primary metal but only 8 million are recycled).

These figures amount to 1 to 2 grams, per member of the entire world population, being disposed of into the environment each year. To cast this figure into perspective, it is noteworthy that "normal" plasma concentrations of aluminium are in the region 10-20 μg dm^{-3} illustrating nature's adept use of chemical speciation, and thus reducing retention and deposition to a minimum. These figures are even more remarkable when one recalls that the aluminium disposal is not spread evenly over the Earth's surface.

It is paradoxical that, because there is such a large amount of aluminium wherever we look, to analyse for rather low concentrations of aluminium species in our food or drink requires exceedingly careful work and the provision of a clean room. Indeed, it has been noted that monitoring such levels in the presence of a high aluminium background is akin to measuring negative concentrations of the element!

This difficulty in measuring micro-concentrations in a mega-concentration environment has partly been responsible for the late discovery of aluminium as a health threat which was only revealed in 1970 when the now well-known link between high aluminium levels in tap water used for renal dialysis equipment was linked with the accumulation of the element in human brain tissue, and possibly with dialysis dementia.

3 BIOAVAILABILITY AND METABOLISM

The scheme shown in Figure 1 gives the various routes through which aluminium can enter into humans from the environment, from our diet, from food and drink additives, from health- and medi-care agents, and from the administration of aluminium as an antacid, antidiarrheal, or as an antiphosphate absorber from the gastrointestinal tract.

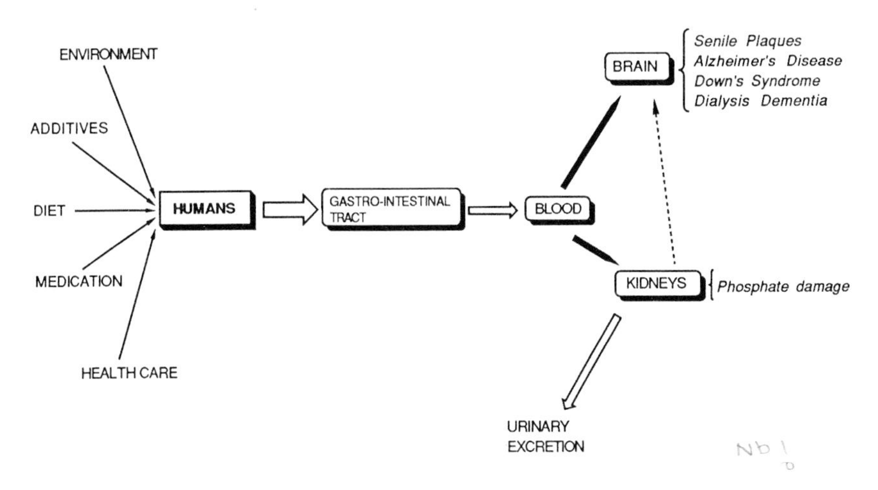

Figure 1: ALUMINIUM PATHWAYS IN HUMANS

Aluminium represents a classical example used by nutritionists to emphasise the differences between intake and uptake. The processes of evolution have conspired in order to preclude aluminium from humans. First, aluminium ores in the geosphere have the element very firmly bound in the form of silicate. Secondly, refined aluminium becomes coated with an inert layer of oxide, making it resistant to corrosion, and a very welcome addition to our technological society. Thirdly, aluminium ions in our diet are completely non-bioavailable from the small intestine as the aquated charged ions are not able to penetrate the lipid protein membranes of the duodenal mucosa and thus pass on into the bloodstream. Those metal ions which are not present in cationic form may be complexed with low molecular mass ligands such as anions of carboxylic acids or amino-acids. Finally, if small amounts of aluminium do enter into the blood plasma, they tend not to be passed through to tissues where they may be retained and alter normal biochemistry but, rather, such charged ions are rapidly excreted by the normal renal mechanisms.

However, when one or more of these safety traps are circumvented by, for example, injecting fluids directly into the bloodstream, by renal insufficiency, or by aluminium present in dust entering the lungs, the aluminium may represent a threat to our normal metabolic processes. Even here, chemical speciation studies have shown that, under physiological conditions, the solubility and charge characteristics of the element are such that any threat remains low.

4 DISCUSSION

The following papers in this monograph illustrate different aspects of aluminium *in vivo* . Professor Edwardson discusses the build-up of aluminium in the brain whilst Dr.Martyn points out a possible link between aluminium levels in drinking water and Alzheimer's Disease. Dr.Stewart comments upon the toxicity to individuals with chronic renal disease. Professor Birchall discusses the bio-inorganic chemistry of aluminium, paying particular emphasis to its interactions with silicic acid, $Si(OH)_4$, phosphate, and even naturally occurring phosphate-containing lipids. He suggests that silicate may be a safety device to complex aluminium and *vice versa* . Dr.Delves describes the great difficulties in monitoring aluminium levels in a variety of samples and Dr.Sherlock shows that the greater the precision of the analytical equipment and techniques used, the lower the real level of aluminium actually found in the wide range of foodstuffs analysed over several decades. Dr.Severus lists the many ways in which aluminium is extending the usefulness of safe storage and transport of food. These authors draw their own conclusions but there are certain themes which reoccur and which are worth emphasising in this introductory section.

- Aluminium is ubiquitious and, therefore, if one is looking for the cause of a physiological condition, it is tempting to "correlate" the incidence with the presence of aluminium. In foodstuffs, aluminium levels of 10 mg per kg are rarely exceeded and, given the very low uptake of the element, these probably represent safe levels. However, over prolonged periods, "a small leak can sink a big ship" and so we must conclude that it is essential for analytical laboratories serving a broad spectrum of professions to be able to analyse for aluminium, even in the presence of high background levels of aluminium.

- The purity of solutions used for parenteral nutrition, of the water used in renal dialysis, and of the components used in preparing bottle feeds for premature babies, ought to be checked out in terms of aluminium, silica, and possibly other trace elements on a routine basis.

- There appears to be a strong need to develop an orally acceptable non-aluminium-based phosphate binder for treating renal disease patients.

- All areas of aluminium biochemistry would benefit from a greater knowledge of its speciation and how other administered agents, such as pharmaceuticals or food additives, affect the speciation. Only through speciation data can the differences between uptake and intake be quantified and optimised. One area meriting particular attention is the need to gain a better understanding of aluminium uptake via the gastrointestinal tract, even though this route represents a very effective barrier.

- A high level of quality control of the data used for publications involving aluminium is essential. This is especially so when the data can cause public alarm such as occurred recently with the unfounded scares concerning aluminium in tea and/or the dissolution of aluminium cooking utensils in the presence of fluoride.

The on-going survey of aluminium containing food additives being carried out by the Ministry of Agriculture, Fisheries & Food will contribute more information to speciation studies and to our general knowledge of aluminium in our diets.

Aluminium Toxicity in Individuals with Chronic Renal Disease

W.K. Stewart

DEPARTMENT OF MEDICINE, UNIVERSITY OF DUNDEE, NINEWELLS HOSPITAL AND MEDICAL SCHOOL, DUNDEE DD1 9SY, UK

This narrative describes the situation whereby, for some four or five decades, physicians, with the best of intentions, have been inadvertently producing ill-effects in their patients with aluminium. I ask you to judge the consequences in the light of the way in which this situation developed.

1 PERCEPTION OF ALUMINIUM SYNDROME

In the early 1960s when long-term artificial kidney treatment began it utilised very basic equipment, and pharmaceutical grade chemicals weighed-out on the spot and mixed in tap water. People were reasonably pleased with the immediate results because without this treatment the patients died within a few weeks and by these simple means they were kept alive. Even so, an impressive catalogue of disparate ailments was apparent even then. The idea that aluminium might be poisonous in this context did not become widely current until after 1970.[1] Familiarity with aluminium hydroxide therapy for both gastrointestinal upset and reducing absorption of ingested phosphate bred a kind of complacency. However, a syndrome (i.e. a stereotyped complex of symptoms and clinical signs) began to be recognised at the bedside in dialysed patients. It was apparently sporadic in occurrence but nevertheless dramatic in its manifestation.

Recognition of Encephalopathy

Initially the affected patient manifested a curious defect of speech, with facial grimacing, and irregular jerking of arms and legs. Eventually epileptic-like seizures developed and progressive dementia led to death in virtually everyone affected, all inside 12 months. This was accompanied by characteristic electroencephalographic (EEG) changes. The "dialysis dementia syndrome" was recognised as a new illness and in 1976 Alfrey et al in the U.S.A. showed an association with accumulating aluminium in the brain.[2] A survey[3] of total European experience identified 150 such patients although many more patients remained unaffected. The great majority of the affected patients had been exposed to dialysis fluid made with untreated or only softened tap water. Only 6% had been dialysed with fluid made from water which was deionised or treated by reverse osmosis. A later survey[4] of American experience over 8 years found 55 patients with encephalopathy (syn. dementia) out of 1380 patients on dialysis, i.e. around 4% overall, with an incidence in one unit as low as 0.2% and in another approaching 15%. In both European and U.S.A. dialysis centres the outstanding correlation was with a high aluminium content in dialysis fluid. In Scotland[5] the patients dialysing within Central Glasgow were free of dialysis dementia, but there were 13 cases scattered around the outskirts of Glasgow and the distribution corresponded with the areas of water supply of known high aluminium content.

It should be noted that not everyone exposed developed evident dementia. Individuals varied in their response to exposure. In Dundee, which has water of low aluminium content (<0.4 μmol l^{-1} for 80% of time) there has been only one dementia-affected patient, and this was associated with a high oral intake of aluminium. Her plasma aluminium content was high (12 - 14 μmol l^{-1}, compared with a normal reference range of <1). Her cerebro-spinal fluid aluminium was raised, and post-mortem examination of the brain showed a high content of aluminium (9-34 μg g^{-1} dry weight, compared with a normal upper limit of <4).

Recognition of Bone Disorder

A complex bone-softening state usually accompanies dialysis-treated chronic renal failure. It is a mixture of 3 independent processes, of which two are predominant. These are:- (1) defective mineralisation of pre-bone collagen, termed osteomalacia, and (2) osteitis fibrosa due to hyperparathyroidism. The rare third process is osteosclerosis. Osteomalacia is ordinarily thought of as being due to deficiency of vitamin D. Gradually it was realised that two out of every three patients with dialysis dementia also had bone disease with crippling bone pain, bone bending and spontaneous multiple fractures. This bone condition was a special variant of osteomalacia, which was not curable by vitamin D and seemed related to the presence of aluminium.[6] In fact, Newcastle-upon-Tyne was once so renowned for the occurrence of this disease in its dialysis patients that it was termed "Newcastle bone disease". Today the belief is that dialysis encephalopathy or dementia and this particular fracturing variant of osteomalacia both represent the effects of accumulating tissue aluminium.

Aluminium analysis by both neutron activation (NAA) and atomic absorption spectrophotometry (AAS) of bone biopsies from apparently-well Dundee patients on dialysis showed a raised aluminium content compared with normals, even when no clinical symptoms were noticeable. Bone biopsies treated with special stains (e.g. aluminon), though not entirely specific, can give some information on the extent of aluminium accumulation at the calcification front, where aluminium appears to be characteristically deposited.[7] This deposition may well be the reason why the osteoid tissue is not calcified. Tetracycline staining in vivo, with two doses separated by a 10-day interval, can indicate the activity of calcification, by highlighting the distance between the ossification front 10 days previously and at the time of the second dose. Patients with demonstrable aluminium at the ossification front show no new tetracycline-stained advance of calcification in 10 days. After treatment with an aluminium chelating agent such as desferrioxamine, new calcified bone formation can be demonstrated as fresh calcium deposition using the 10-day tetracycline staining test.[8] However, this procedure requires quite expensive histopathological facilities which are not generally available throughout the NHS.

Dialysis Fluid Aluminium

With less than 0.4 $\mu mol\ l^{-1}$ of aluminium in the dialysis fluid and less than 1 $\mu mol\ l^{-1}$ in the blood plasma, transfer of aluminium into the patient should be minimal (Figure 1a). The molecular form of the aluminium at the pH of dialysis fluid, which also contains saline, bicarbonate and other compounds, is complex - at least to clinicians. However, if the aluminium content of dialysis fluid is appreciably more than 0.4 $\mu mol\ l^{-1}$ (Figure 1b), there is a continuous transfer across the dialyser membrane into the patient. That transfer is almost unimpeded over the period of each dialysis session because the aluminium which has moved across is rapidly and specifically bound by a transport protein normally present in blood plasma, transferrin, so promoting further inwards diffusion. The patient therefore progressively takes in a load of aluminium thrice weekly with each dialysis. The result is that the plasma transferrin is increasingly saturated with aluminium, total plasma aluminium increases and there appears to be a concomitant progressive transfer from transferrin into tissue depots.

Dialysis fluid is now treated so as to eliminate or at least reduce to the minimum the aluminium content. The main effort is applied to the tap water. Dialysis fluid is made by proportionating 34 parts of mains water and one part of commercial dialysis fluid concentrate. In the dialysis unit, the mains water is subjected to a preliminary filtration to get rid of coarse, mud-like materials, then softened, charcoal-filtered for the final removal of particulate material and lastly subjected to reverse osmosis before proportionate-mixing with the concentrate. Reverse osmosis of the tap water greatly reduces the aluminium content in the product dialysis fluid. Dialysis fluid concentrate is nowadays prepared in bulk by only two or three manufacturers in this country and it is accepted that the ingredients must be as free from aluminium as it is possible to make them. The arbitrary desiderata, which have been arrived at largely empirically, are that the mains water content of aluminium should preferably be less than 30 $\mu g\ l^{-1}$ (1.1 $\mu mol\ l^{-1}$), ideally less than 10 $\mu g\ l^{-1}$ (0.4 $\mu mol\ l^{-1}$) and that the final total aluminium in product dialysis fluid[9] should never exceed 30 $\mu g\ l^{-1}$, and ideally should be less than 10 $\mu g\ l^{-1}$.

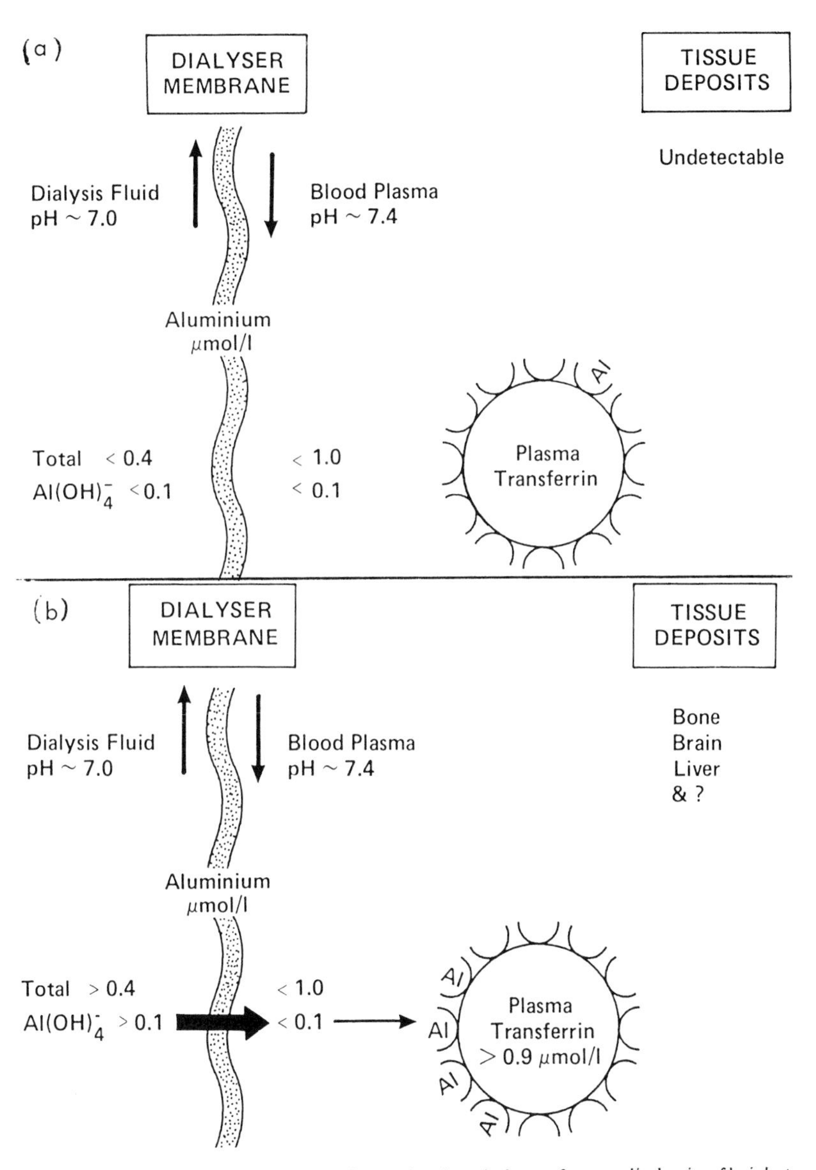

Figure 1. Mechanism of transfer of aluminium from dialysis fluid to patient during haemodialysis using dialysis fluid (a) with low concentrations of aluminium (<0.4 μmol l^{-1}) and (b) with high concentrations of aluminium (>0.4 μmol l^{-1})

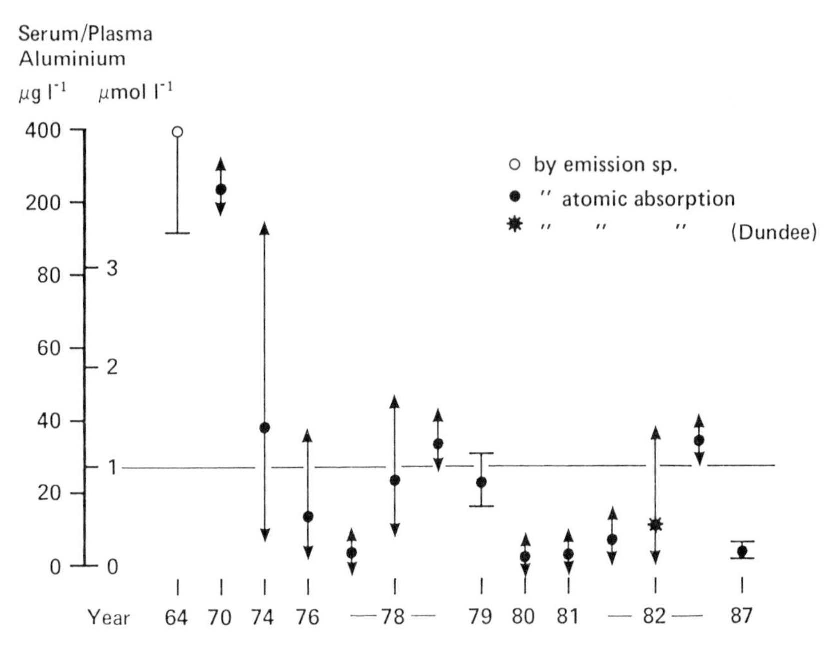

Figure 2. Reported normal reference ranges[10] for serum or plasma aluminium concentrations between 1964 and 1987, showing the variation and general decrease with time (I = SD, ↕ = range)

Clinical Monitoring

The greatest stricture over the years since 1970 has been the sheer difficulty in measuring plasma aluminium. Even today this has to be done in a supra-regional laboratory. Furthermore, normal ranges have gradually changed with time[10] as the few specialised laboratories gained experience with the demanding technique and learned by hard-won experience to eliminate extraneous aluminium (Figure 2). The amounts present in normal plasma are very low but increases seem to be significant biologically and warrant clinical attention. One has to admit that the plasma aluminium level on any one day represents only an approximate index of recent exposure to and absorption of aluminium and is not in any sense a measure of total "body burden".[11] Recently acquired aluminium is progressively transferred from the plasma to tissue. One of the questions we cannot yet answer in a living patient is 'what is the total tissue burden - what is the individual organ content of aluminium'?

Regular serial monitoring of blood levels is essential,[12] with all the expense that that involves (Figure 3). Only one current patient in Dundee has consistently maintained high plasma aluminium levels

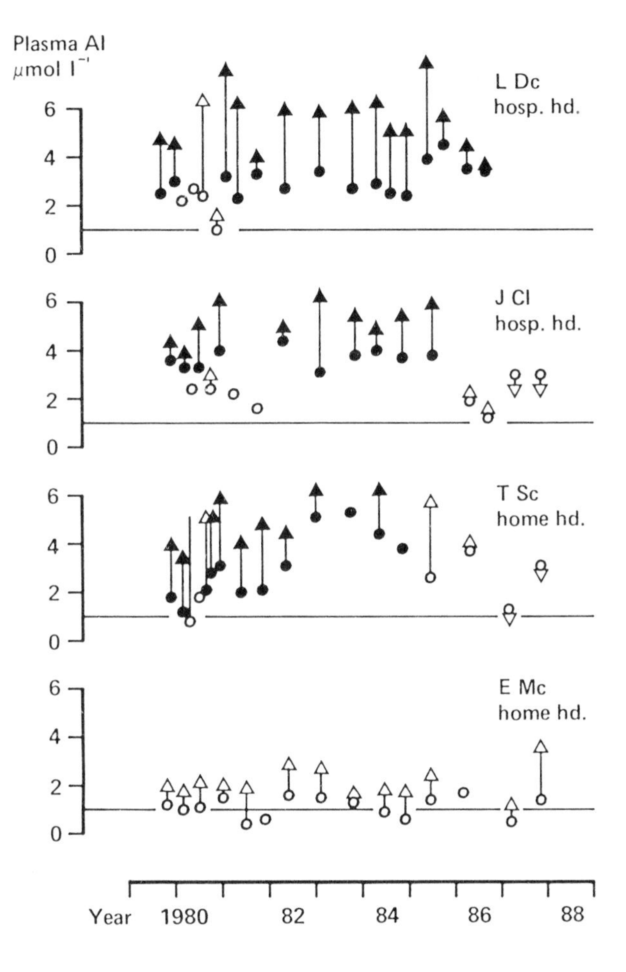

Figure 3. Serial monitoring of plasma aluminium concentrations over 8 years in 4 patients on maintenance haemodialysis (predialysis - o not taking oral aluminium, ● taking oral aluminium, and post-dialysis - Δ not taking oral aluminium, ▲ taking oral aluminium).

(range 5 - 8 µmol l^{-1}), despite all attempts to minimise exposure and, although there are no overt symptoms of aluminium toxicity, he may well be suffering from some of the less overt conditions of overload. One never knows when some item of dialysis equipment is failing or when the environmental situation as regards aluminium exposure for each patient is changing unnoticed. Seasonally the mains water content of aluminium alters as does the pH. Post-dialysis plasma aluminium levels in our patients are usually higher than pre-dialysis levels, due largely to ultrafiltration which concentrates all protein plasma, and such fluctuations could also conceal ingress of aluminium and transfers between tissue depots and dialysis fluid. Those patients taking oral aluminium hydroxide have demonstrably higher plasma aluminium. This points to the intestinal absorption problem which has gradually gained importance as dialysis fluid aluminium exposure has lessened.

Desferrioxamine (DFO) can be used clinically as a semi-quantitative test of tissue burden of aluminium, as well as a means of treating aluminium overload.[13] It is a compound of biological origin, produced as an iron chelate within a fungus. The pharmaceutical industry can isolate the free ligand which then shows a marked affinity for iron and in lesser degree for aluminium, and to only a slight extent for calcium. It is possible to give this material parenterally and as it circulates in the plasma it appears to attract iron and aluminium from the tissue depots. Chelated iron and aluminium then circulate in the blood in soluble form attached to the DFO and can be removed intact through the dialyser. This method can be used to treat both iron and, in the present context, aluminium poisoning but it is not entirely free of toxic hazard. DFO can for example produce lens cataracts and it is not a specific agent which exerts its effect solely on either aluminium or iron. One cannot remove one without the other. Nevertheless DFO infusion has been used as a test for the tissue burden of aluminium.[14] The effect of giving 0.5 g DFO post-dialysis is to cause an elevation of the total serum aluminium from 3 to 7 μmol l^{-1} by the time of the next haemodialysis session and this increment is rapidly eliminated via the dialyser in the course of that second dialysis. The extent to which the serum aluminium level rises over the two days after dosage is said to be an index of the stores of aluminium. Unfortunately, like bone biopsies, such tests for the body burden of aluminium are imperfect, cumbersome, expensive, and therefore, not frequently done. In routine terms, we have to make do with serial serum or plasma aluminium levels, so intrinsic difficulties remain in evaluating the extent of an individual's aluminium load.

Alimentary Absorption

The other major problem is the extent of absorption of ingested aluminium compounds.[15] All the time during which awareness of aluminium-related ill effects and its possible relation to dialysis fluid content was growing, many clinicians were still giving their patients oral aluminium-containing drugs in the belief that the aluminium in compounds administered orally was not absorbed and that the transfer from dialysis fluid was the main or only source of the encephalopathy/ osteodystrophy problem. Berlyne, to his great credit, claimed early on that orally-administered aluminium could be absorbed and was potentially toxic.[1]

The paramount biological feature of aluminium is how well it is kept out of the human or mammalian body under normal conditions by the function of the alimentary tract (Figure 4). Normally a relatively

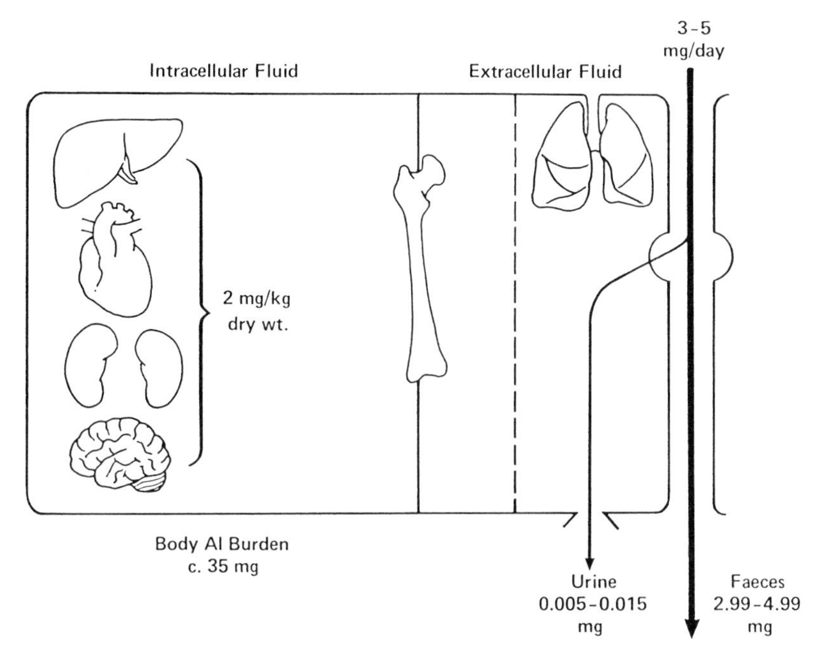

Figure 4. Aluminium economy of intake and excretion under normal conditions, with all but a small fraction of normal daily intake being excreted in the faeces.

small amount of aluminium enters the mouth and the great majority passes through the gut and out into the faeces, unabsorbed. The tiny fraction which is absorbed, mechanism unknown, is excreted by the normal kidney. The therapeutic impasse arises indirectly, because patients with advanced renal failure progressively calcify visceral soft tissues, arteries, joint capsules and the cornea, due to the unwanted deposition of calcium phosphate. The phosphate component is of dietary origin, arising from the oxidation of proteins. It is very desirable to avoid this deposition of calcium phosphate and the means adopted is to reduce phosphate absorption from the gut by administering aluminium hydroxide. This way of reducing phosphate absorption began in 1941,[16] purely empirically, and has been used ever since. Hyperphosphataemia can be alleviated indirectly by complexing the dietary phosphate inside the gut lumen so that it is excreted in the faeces. Aluminium hydroxide also has a well known dyspepsia-relieving, antacid action and it is widely used for treating duodenal ulcers. Patients with renal failure on aluminium hydroxide do not generally need an antacid for indigestion and are taking it purely to complex and remove inorganic phosphate via the gut. Unfortunately, the quantities needed for sufficient phosphate-binding have to be much greater than those used conventionally for

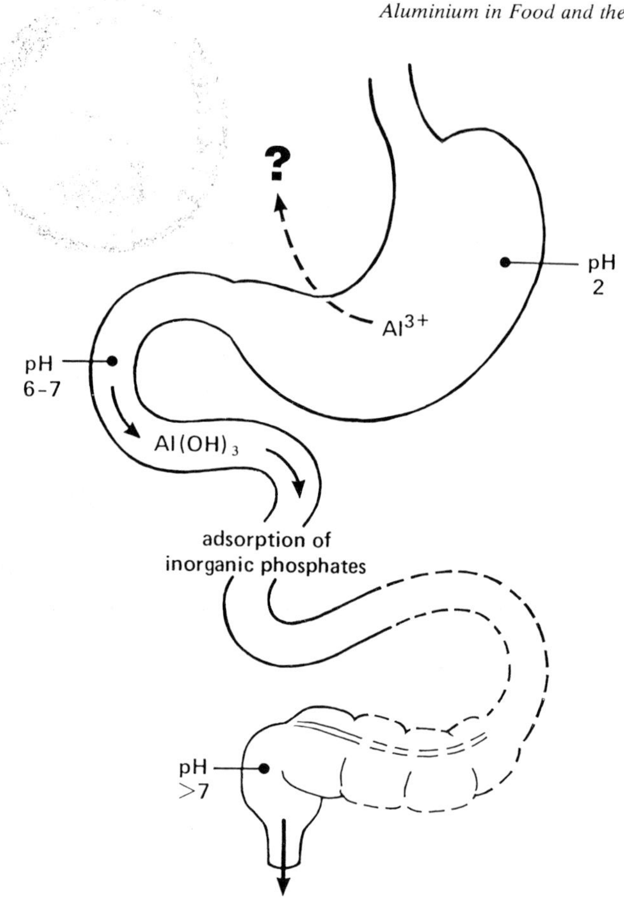

Figure 5. Sequestration of inorganic phosphate in the gut by precipitated aluminium hydroxide at pH 6 and possible route of absorption of free ionic aluminium at pH 2 in the stomach.

dyspepsia. Patients with renal failure may be given two or three grams of aluminium per day in aluminium hydroxide form over many months - enough that is to reduce the serum phosphate ideally to almost normal. But, by solving the excess phosphate problem, we go on to create an aluminium problem.

Depending on ambient pH (Figure 5) the chemical form and solubility of aluminium compounds change throughout the anatomical course of the gut. Various forms of aluminium hydroxide or aluminate can exist within the gut lumen. Aluminium hydroxide at acid pH is solubilised and appears to form a free trivalent ion. At pH 5-6 it is least soluble, but with the soluble part in several hydrated forms and at

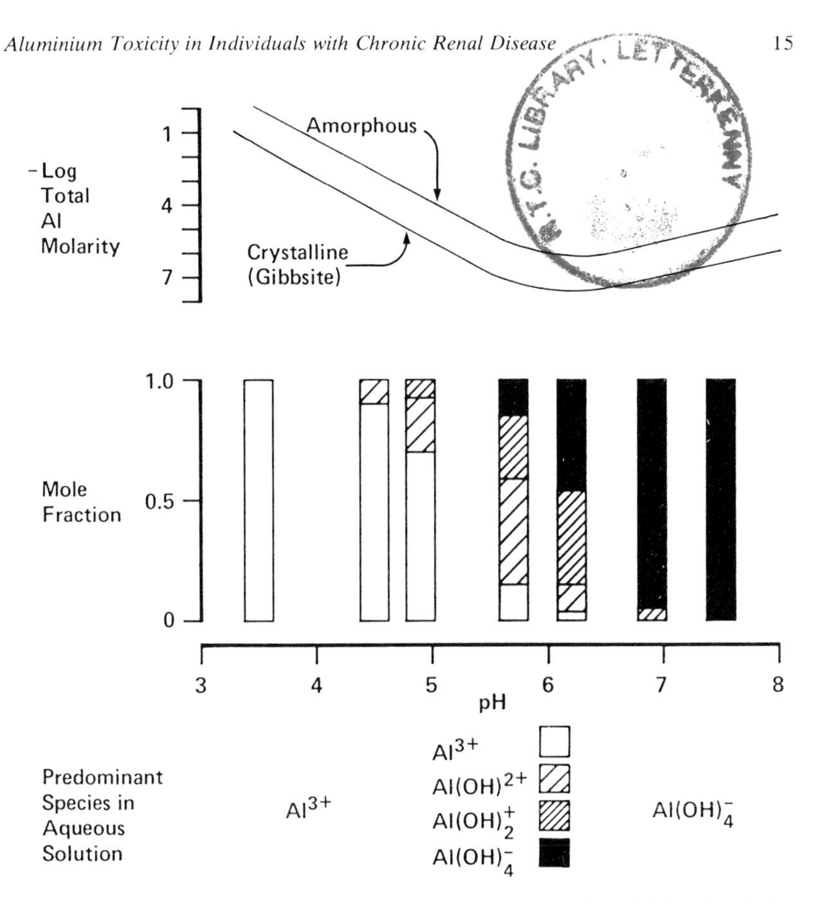

Figure 6. Solubility curves and molecular forms of soluble aluminium hydroxide at different pH values (Redrawn from data of Martin 1986).[17]

pH >6 to 7 it exists as a complex hydroxide (Figure 6).[17] All this goes on inside the gut when clinicians prescribed these large quantities to be ingested along with the varied constituents of food intake. It is now certain that some of the therapeutic aluminium is absorbed.[15] Where exactly it's being absorbed and in what molecular form it is being absorbed is still obscure.

Additional Clinical Concomitants

Another recently recognised aspect of aluminium toxicity, in addition to the recognised effects on brain and/or bone, is anaemia which is probably due to inhibition in the bone marrow of the haemoglobin synthetic process. It is claimed that anaemia, characteristically but not inevitably hypochromic and microcytic in morphology, may be the sole evident sign of slowly accumulating or protracted aluminium poisoning[18]. It has also been suggested on much less evidence that smaller amounts of aluminium accumulation may lead to psychomotor defects in the brain which are much less obvious than the flagrant dialysis dementia caused by aluminium. The full extent of the clinical syndromes associated with the aluminium accumulation has not yet been "plumbed", if one dare use that term. The "peak of the iceberg" has been recognised but there may be other features of toxicity yet to be discovered which are below clinical recognition at the moment. One must remember that the realisation that the more obvious acknowledged abnormalities were due to aluminium was initially based on supposition by gifted clinical observers. The possibly harmful effects of less extensive aluminium loads are not yet known. It is unfortunate that aluminium measurements have been so sparsely available to clinicians.

Other Extraneous Sources of Aluminium

It has been found that a variety of parenterally administered fluids, such as plasma exchange fluids, plasma protein fraction fluids, peritoneal dialysis fluids and enteral formula feeds, are all occasionally and unexpectedly contaminated with aluminium from filters or other parts of the manufacturing process. It is evident that constant vigilance for inadvertant aluminium contamination is now necessary in all treatment areas, even outwith the area of renal impairment.

2 COMPLEXITY OF CHRONIC RENAL FAILURE

Trace elements in renal failure have not always received the attention they deserve. There are a number of trace elements which are widely admitted to be either essential and/or potentially toxic in excess, and these are certainly disturbed in chronic renal failure, especially in patients on maintenance haemodialysis. Fluorine, silicon and chromium are increased in the plasma of these patients, and nickel, zinc and selenium are found to be reduced. The effects produced by, or associated with, this complex disturbance of the normal pattern are pure conjecture at the present time. What part do the other elements play in relation to aluminium? For example, does the concurrent reduction in selenium make a contribution to the aluminium excess syndromes? No one knows. We do not really know whether or not the syndrome now described as aluminium-related dialysis osteomalacia represents some more complex disturbance which is shared by aluminium and some or even all of the other elements. Currently, it seems as through aluminium plays the main part, but the role played by the other trace elements is not yet determined since sufficient exploratory or experimental measurements have not been done and not enough people have the analytical means to explore the clinical situation.

3 PREDICTION OF BIOLOGICAL PROPERTIES AND TOXICITY

The patterns of the Periodic Table led to useful rationalisation of physical properties and chemical reactivities. It seems a great pity that there is not some corresponding "biological" periodic table which meaningfully interrelates the elements in terms of their biological roles. One or two people have attempted to tackle this.[19] In Fearon's two Tables, beryllium was excluded as a biologically unnecessary element but aluminium was included amongst those elements which are acceptable biologically and which might even have a useful biological role. So predictions for aluminium on that basis were wrong. No examples of an enzyme which requires aluminium as a catalyst have apparently been found. A great many trace elements have their role in relation to enzyme activity but not, significantly, aluminium.

Looking backwards it's possible to see many reasons why aluminium might have been suspected as potentially toxic long before 1970. Everybody is taught that there is an enormous amount of aluminium present in the earth's crust and that prevalence was once argued as the reason why man, or even the animal domain as a whole, didn't need to worry about it as their progenitors had been exposed to aluminium from the beginning of time and it had created no recognised problem. Indeed, this was still being written in the late 1970s. Sea water, certainly, contains relatively tiny amounts of aluminium, something like 1 $\mu g\ l^{-1}$. The tissue fluids of the mammalian body, blood plasma for example, have often been likened in some ways to sea water, perhaps representing an earlier primordial aquatic origin. There is little more aluminium present in human plasma than in present day sea water. The inference to be drawn is that aluminium is kept exteriorised by higher animals.

In conclusion, it is dangerous to overpower or circumvent, by feeding or well-intentioned therapy, the alimentary mechanism which protects us from soluble aluminium. Patients with renal failure are the paradigm. They are at risk when the kidneys are diseased, urinary excretion is diminished, and especially when coupled with that, they are exposed to extra aluminium by either or both parenteral and oral routes.

Aluminium has been both indicted and incriminated as harmful, perhaps in concert with disturbances of other trace elements. It is a salutary lesson to see that such low concentrations can exert such devastating effects. Physicians were perhaps slow in perceiving the clinical pattern which resulted from excess aluminium. Trace elements are generally thought of as being potentially valuable micronutrients. The dominant idea has long been that we mustn't do without them because we might be ill through deficiency. We must begin to think of trace elements also as possibly micropoisons when in excess, with the related clinical syndromes yet to be recognised. The inorganic chemistry of human disease needs as much appraisal in depth as organic biochemistry already receives in today's world.

REFERENCES

1. G.M. Berlyne, J. Ben Ari, D. Pest, J. Weinberger, M. Stern, G.R. Gilmore and R. Levine, Lancet, 1970, 2, 494.
2. A.C. Alfrey, G.R. LeGendre and W.D. Kaehny, New Eng. J. Med., 1976, 294, 184.
3. A.J. Wing, F.P. Brunner, H. Boynger, C. Chantler, R.A. Donckerwolcke, H.J. Gurland, C. Jacobs, P. Kramer and N.H. Selwood, Lancet, 1980, 1, 190.
4. M.T. Schreeder, M.S. Favero, J.R. Hughes, N.J. Petersen, P.H. Bennett and J.E. Maynard, J. Chronic Dis., 1983, 36, 581.
5. H.L. Elliott, F. Dryburgh, G.S. Fell, S. Sabet and A.I. MacDougall, Brit. Med. J., 1978, 2, 1101.
6. M.K. Ward, T.G. Feest, H.A. Ellis, I.S. Parkinson and D.N.S. Kerr, Lancet, 1978, 1, 841.
7. P.S. Smith and J. McClure, J. Clin. Path., 1982, 35, 1283.
8. S.M. Ott, D.L. Andress, H.G. Nebeker, D.S. Milliner, N.A. Maloney, J.W. Coburn and D.J. Sherrard. Kidney Int., 1986, 29, Suppl. 18, S-108.
9. European Economic Community Resolution 86/C - 184.04. Official J. Eur. Comm. C184 16 July 1986.
10. R. Cornelis and P. Schutyser, Contrib. Nephrol., 1984, 38, 1.
11. N.W. Boyce, S.R. Holdsworth, N.M. Thomson and R.C. Atkins, Nephron, 1987, 45, 164.
12. R.J. Winney, J.F. Cowie and J.S. Robson, Kidney Int., 1986, 29, Suppl. 18, S-91.
13. P. Ackrill and J.P. Day, Contrib. Nephrol., 1984, 38, 78.
14. D.S. Milliner, H.G. Nebeker, S.M. Ott, D.L. Andress, D.J. Sherrard, A.C. Alfrey, E.A. Slatopolsky and J.W. Coburn. Ann. Intern. Med., 1984, 101, 775.
15. L.W. Fleming, W.K. Stewart, G.S. Fell and D.J. Halls, Clin. Nephrol., 1982, 17, 222.
16. G.B. Fauley, S. Freeman, A.C. Ivy, A.J. Atkinson and H.S. Wigodsky, Arch. Intern. Med., 1941, 67, 563.

17. R.B. Martin, Clin. Chem., 1986, 32, 1797.

18. R.D. Swartz, Proc. E.D.T.A., 1985, 22, 101.

19. W.R. Fearon, Sci. Proc. R. Dublin Soc., 1951, 25, 235.

Aluminium and the Pathogenesis of Neurodegenerative Disorders

J.A. Edwardson, A.E. Oakley, R.G.L. Pullen, F.K. McArthur, C.M. Morris, G.A. Taylor, and J.M. Candy

MRC NEUROCHEMICAL PATHOLOGY UNIT, NEWCASTLE GENERAL HOSPITAL, NEWCASTLE UPON TYNE NE4 6BE, UK

1 INTRODUCTION

Neurodegenerative disorders are conditions in which there is slow, progressive loss of specific populations of neurones in the central nervous system, giving rise to characteristic functional deficits and eventual death. They include Alzheimer's disease, Parkinson's disease, motor neurone disease, Huntington's disease and a large number of less common conditions, each with its own pattern of selective neuronal vulnerability and clinical impairment. Certain neurodegenerative disorders such as Huntington's disease and some familial forms of early onset Alzheimer's disease are inherited in an autosomal dominant fashion and have been shown in genetic linkage studies to associate with chromosomal markers indicating a single gene defect. However, many neurodegenerative conditions are sporadic in their occurrence and there is increasing evidence that these may result from environmental factors such as toxins, infectious agents or traumatic injury which selectively damage certain types of neurone.[1] Examples include parkinsonism induced by methylphenyltetrahydropyridine, the post-poliomyelitis syndrome following infection with polio virus and dementia pugilistica resulting from traumatic head injury in boxers. The damage caused by such environmental factors may occur long before the appearance of clinical symptoms; these subsequently develop because of further age-related attrition of neurones from the affected systems and the associated failure of compensatory mechanisms.

Aluminium has been implicated in the pathogenesis of four neurodegenerative conditions. There have been rare reports of encephalopathy caused by the massive inhalation of aluminium dust in an industrial environment.[2] Aluminium has been implicated as a major factor in the progressive encephalopathy which affected many patients on haemodialysis for chronic renal failure,[3] a condition described elsewhere in this volume. It has been reported that the intracellular content of aluminium is raised in nerve-cells which contain neurofibrillary tangles in the amyotrophic lateral sclerosis-parkinsonism-dementia complex (ALS-PD) of Guam.[4] This is a severe neurodegenerative condition found in high incidence foci on Guam and other islands of the Western Pacific, which is characterized by a combination of neurological symptoms including muscular weakness and paralysis due to loss of motor neurones, parkinsonian movement disorder and dementia. Neurofibrillary tangles are a major neuropathological lesion in the ALS-PD complex. It has been proposed that the intracellular accumulation of aluminium and also calcium in tangle-bearing neurones is secondary to a defect in brain mineral metabolism resulting from a chronic nutritional deficiency of calcium and magnesium which, in turn, gives rise to the increased absorption of toxic metals such as aluminium and manganese.[5] However, it has recently been shown that a toxic amino-acid, β-N-methylamino-l-alanine, is a constituent of the fruit of the false sago palm which was used extensively as a food source and medicine by the populations affected with ALS-PD.[6] This toxin was fed to macaque monkeys and produced neurodegenerative changes in the brain and spinal cord with marked symptoms of parkinsonism and motor impairment. While it is possible that ALS-PD may result from the combined effects of several environmental factors including aluminium, the cause of this disease remains unknown and is unlikely to be determined, owing to the sharp fall in its incidence over recent decades.[5] By far the most important neurodegenerative condition which has been linked to aluminium is Alzheimer's disease.

2 ALZHEIMER'S DISEASE

Alzheimer's disease is the most common neurodegenerative disorder. It is the major cause of senile dementia, a condition which in a moderate to severe form affects approximately 5% of the population aged 70 years or over in the United Kingdom. The chief clinical symptoms are loss of short-term memory and severe, progressive impairment of intellectual functions in general. It is accompanied by neuropathological changes which include neuronal loss and two hallmark diagnostic features - neurofibrillary tangles and senile plaques (Fig 1). Tangles consist

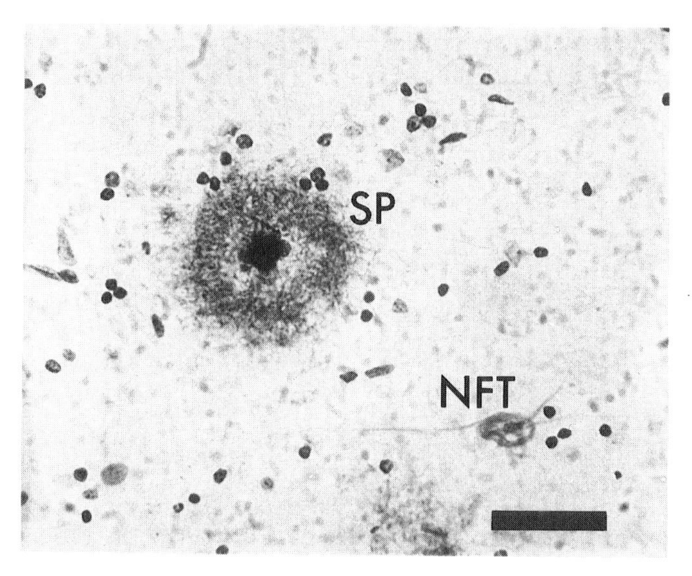

Figure 1. Senile plaque (SP) and neurofibrillary tangle (NFT), the two characteristic neuropathological lesions of Alzheimer's disease. Bar scale = 40 microns.

of intraneuronal masses of filaments with a characteristic paired helical structure.[7] Their chemical composition has not yet been elucidated although a minor component consists of a fragment of tau protein, a molecule involved in regulating the assembly of microtubules.[8] Senile plaques are more complex structures, consisting of abnormal nerve and glial cell processes, often surrounding a central core of fibrillary amyloid protein. The number of senile plaques in the cerebral cortex correlates both with the severity of clinical impairment and the deficit in biochemical markers of cholinergic neurotransmission. Loss of acetylcholine appears to be a major cause of the memory loss in Alzheimer's disease.[9] Three forms of senile plaque occur: 'primitive' amorphous plaques, composed of loose aggregates of amyloid fibrils; 'classical' neuritic plaques in which there is a central core of densely compacted amyloid, surrounded by dystrophic neuronal processes and glial cells; 'compact' plaques which consist solely of a dense amyloid core. These different forms were previously thought to represent a developmental sequence[10] but this has not been confirmed and they may have quite different mechanisms of pathogenesis.

The recent isolation and characterization of the senile plaque amyloid protein represents a major advance in understanding the pathogenesis of Alzheimer's disease. Amyloid deposits are also found around cerebral and meningeal blood vessels - a condition known as congophilic angiopathy. Glenner et al isolated perivascular amyloid from meningeal blood vessels and showed that the major constituent was a 42 amino acid polypeptide.[11] A similar peptide (the A4 or β-amyloid protein) was isolated from senile plaque cores and shown to be derived from a large glycoprotein precursor.[12-15] The gene for this precursor is located on chromosome 21, a finding of considerable interest because Alzheimer-type neuropathological changes invariably occur in individuals with Down's syndrome, or trisomy 21, by the age of 40-50 years. The A4 precursor seems to be a membrane associated protein with part of the hydrophobic transmembrane-spanning sequence forming the subunit of polymeric A4 fibrillary aggregates in amyloid deposition. Abnormal expression or proteolysis of the A4 precursor is believed to be central to the cellular pathology of Alzheimer's disease.

3 ALUMINIUM AND ALZHEIMER'S DISEASE

The possible involvement of aluminium in the pathogenesis of Alzheimer's disease was first proposed following the chance observation that injections of aluminium salts in rabbits and other susceptible animal species produced neurofibrillary tangles, resembling in their light microscopic appearance those seen in man.[16] However, at the ultrastructural level, such experimentally induced tangles are composed of 10nm straight filaments,[17] in contrast to the paired helical filaments which form the tangles in Alzheimer's disease. While this suggests that the two types of tangle may be unrelated, more recent studies have shown that aluminium-induced tangles in rabbits contain abnormally phosphorylated neurofilament proteins[18] similar to those which decorate tangles in man[19], and it is possible that similar mechanisms may be involved in their pathogenesis.

Further interest in an aetiological or pathogenic role for aluminium was stimulated by claims that the aluminium content of the brain is increased in Alzheimer's disease.[20,21] These reports were subsequently contradicted by later studies which failed to find a significant difference between the aluminium content of Alzheimer brains and age-matched control groups, although an age-related increase in the brain content was reported.[22-24] However, there is a significant age-related increase in the burden of Alzheimer-type neuropathological changes in non-demented elderly individuals[25] and studies to-date have not attempted to correlate tissue levels of aluminium with the burden of such neuropathological changes *per se*, rather than with age. This important issue has not been resolved and a recent report claims a substantial increase in the aluminium content of cerebral cortex and hippocampus in a group comprised mainly of presenile cases of Alzheimer's disease, compared with age-matched controls.[26] The relationship between brain tissue levels of aluminium and neuropathological changes is difficult to resolve by gross tissue measurements since aluminium is present at extremely low concentrations and is likely to be present in several cellular compartments, including neurones, glia and vascular endothelial cells. Thus it has proved necessary to use microanalytical imaging techniques which permit the direct mapping of Al and other trace elements associated with neuropathological features.

Using electron microprobe X-ray microanalysis, Perl and Brody reported in 1980 that aluminium was present in the majority of tangle-bearing neurones from the hippocampus of senile cases of Alzheimer's disease, in contrast to adjacent neurones without tangles in the same cases or from non-demented elderly individuals.[27] A recent study has failed to confirm this association but the material investigated consisted of formalin-fixed, iron haemotoxylin and eosin-stained, gold-coated paraffin sections where both loss of bound aluminium and attenuation of X-ray signals is likely to occur.[28]

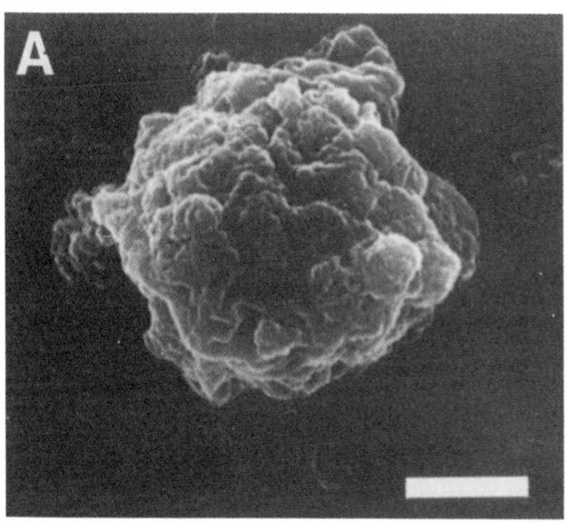

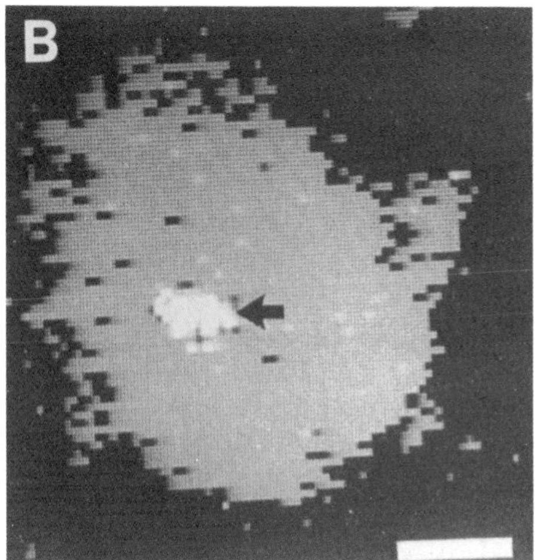

Figure 2. (A) Scanning electron micrograph of isolated senile plaque core. Bar scale = 4 microns. (B) Scanning electron microprobe X-ray analysis of senile plaque in tissue section. The central white core shows aluminium and silicon co-localized in the core, while the surrounding grey area reflects the silver staining of neuritic processes. Bar scale = 20 microns.

In the course of attempts to isolate the amyloid protein from senile plaque cores, we became aware that a significant proportion of isolated plaque core preparations consisted of non-protein material. Investigations using scanning electron microprobe X-ray analysis revealed the presence of focal, co-localized deposits of aluminium and silicon at the centre of the senile plaque core and solid-state nuclear magnetic resonance spectroscopy confirmed the presence of amorphous aluminosilicate.[29-31] Such focal deposits were observed in plaque cores from presenile and senile cases of Alzheimer's disease, non-demented elderly individuals and cases of Down's syndrome.[32] They were evident (Fig 2) not only in isolated cores but also in situ, in tissue sections, and we have confirmed the distribution of both aluminium and silicon using scanning proton energy dispersive X-ray microanalysis[33] and imaging secondary ion mass spectrometry.[34] Preliminary studies suggest that similar deposits of aluminosilicate are not associated with other forms of cerebral amyloid (congophilic angiopathy or scrapie plaques) or systemic amyloid (light chain IgG or serum amyloid A protein) deposition or other neuropathological lesions including corpora amylaceae[34] and Pick bodies.[35] The possibility that aluminium and silicon may have a pathogenic role in plaque formation is raised by the fact that these elements are consistently and specifically co-localized at the centre of the plaque core. A secondary, non-pathogenic deposition of either aluminium or silicon on polymeric protein aggregates would be expected to produce a more variable accumulation throughout the plaque core and to be found in association with other neuropathological features involving fibrillary aggregates. These considerations have led us to investigate the mechanisms by which aluminium and silicon enter the brain and could influence the formation of A4 amyloid deposits.

4 ALUMINIUM AND SILICON UPTAKE BY THE BRAIN

The metabolism of aluminium has been reviewed extensively by Ganrot.[36] There is a marked gastro-intestinal barrier to absorption and rapid urinary excretion removes the small quantities absorbed in normal subjects. Most, if not all aluminium in blood is bound to transferrin, which has a high affinity for this element.[37,38] Aluminium-loaded transferrin is

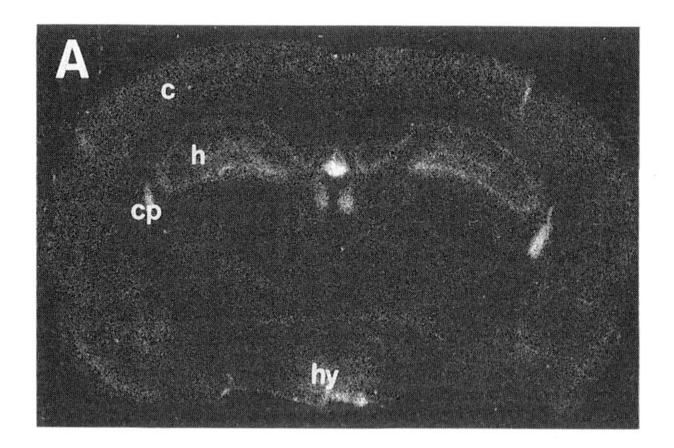

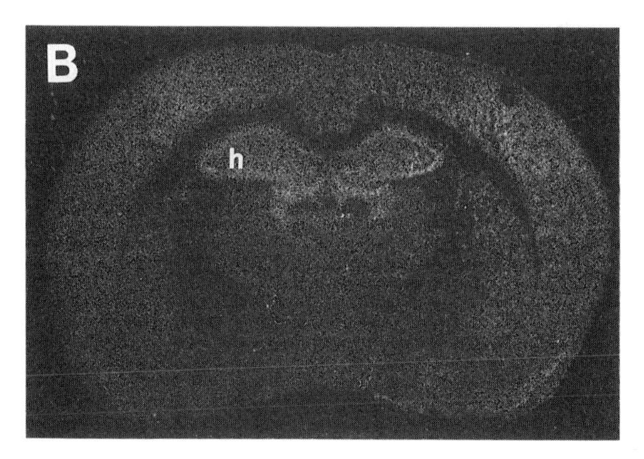

Figure 3. Autoradiographs of rat brain showing (A) the uptake of ^{67}Ga and (B) the distribution of transferrin receptors. Note the high uptake of ^{67}Ga in areas such as the cortex (c) and hippocampus (h), and where there is no blood-brain barrier such as the choroid plexus (cp) or hypothalamus (hyp).

bound and internalized by cultured neuroblastoma cells[39] and the same mechanism would appear to operate for cerebrovascular endothelial cells or cells of the choroid plexus since aluminium can cross the blood-brain barrier. In the absence of a suitable radioactive isotope of aluminium we have used ^{67}Ga as a marker for aluminium transport and shown that the highest levels of uptake occur in the cerebral cortex, hippocampus, septum and amygdala.[40,41] These are areas which contain the highest densities of transferrin receptors and also coincide with the regions which are selectively vulnerable in Alzheimer's disease (Fig 3). Since the density of transferrin receptors in the cerebral cortex remains unaltered in Alzheimer's disease,[42] this system appears to provide a possible mechanism for aluminium entry. The uptake of aluminium into brain via the iron transport system is consistent with observations on senile plaque cores using a scanning proton microprobe where small focal deposits of iron were also shown to be present.[33] If the capillary permeability of aluminium in man is similar to that found for ^{67}Ga (3.6×10^{-6}ml/min/g) in the rat, we have calculated[41] that the brain will accumulate approximately 10μg per annum, a figure consistent with the observed content of the human brain after 70 years.[23] Silicon, unlike aluminium does not appear to be bound to a protein carrier in blood and probably is present as silicic acid. We have used ^{68}Ge as a probe for silicon entry into rat brain and shown that there is a high capillary permeability compared to ^{67}Ga, with bi-directional passage and a relatively even regional distribution, similar to the uptake of other small, non-protein bound solutes.[41]

5 CHANGES IN CHRONIC RENAL DIALYSIS

The association of Al and Si with plaques in Alzheimer's disease, and the existence of transport mechanisms for their uptake in brains of normal subjects raise the possibility that increased exposure to these elements *per se* can result in neurodegenerative changes. In order to test this hypothesis we have examined the brains of a group of patients with chronic renal failure.[43] Such patients are unable to excrete aluminium or silicon adequately and are frequently treated with relatively large doses of aluminium-containing compounds to reduce phosphate absorption. As Table 1 shows, there was a significant increase in the brain content of both Al and Si, as determined in cortical samples using graphite furnace atomic absorption spectrometry (GFAAS). Measurements using inductively-coupled plasma mass spectrometry (ICPMS) suggest that this increase is specific for these elements, although such bulk analyses may obscure significant differences in distribution at the cellular level.

Table 1 Gross tissue content of elements in cerebral cortex as determined by ICPMS and *GFAAS

Element	µg/g dry weight tissue	
	mean controls (range)	mean dialysis (range)
*Al	2.6 (1.0 - 7.8)	6.3 (1.0 - 17.7)
*Si	3.8 (2.3 - 6.0)	6.1 (2.4 - 14.4)
Mg	475 (429 - 540)	491 (458 - 549)
Mn	2.0 (1.2 - 3.2)	2.0 (1.2 - 3.3)
Fe	246 (227 - 298)	232 (174 - 320)
Ni	0.9 (0 - 1.5)	1.0 (0 - 1.9)
Cu	20.5 (17.2 - 24.5)	15.3 (12.5 - 17.2)
Zn	43.4 (36.8 - 50.6)	45.8 (40.7 . 52.7)

N=15 for both groups for determinations by GFAAS.
N=10 for both groups for determinations by ICPMS.
Levels of Al were found to be similar using both procedures but it was not possible to measure Si by ICPMS due to mass interference.

Immunocytochemical staining with antisera raised against synthetic A4 amyloid peptide revealed high densities of senile plaques in the cerebral cortex (Fig 4) in 5 out of 15 dialysis cases. Plaques are not usually seen in individuals below the age of 55 years, except in pre-senile Alzheimer's disease and Down's syndrome (CL Masters, personal communication) but immature plaques were present in 2 out of the 6 dialysis cases aged 55 years or younger. The presence of diffuse extracellular deposits of amyloid fibrils associated with the plaques in these dialysis cases was confirmed using electron microscopy on silver stained sections. Mature plaques containing a dense amyloid core were rarely seen. Neurofibrillary tangles were not present in the cerebral cortex in any of the dialysis cases, even after immunocytochemical staining with antisera against paired helical filaments or Alz 50 protein[44], which readily showed such changes in Alzheimer brain.

There was no correlation between the gross brain tissue levels of aluminium and the occurrence of immature senile plaques, although levels of silicon tended to be higher in the dialysis cases with plaques. Tissue levels of aluminium and silicon were below the limit of detection using scanning electron microprobe X-ray microanalysis but SIMS revealed a

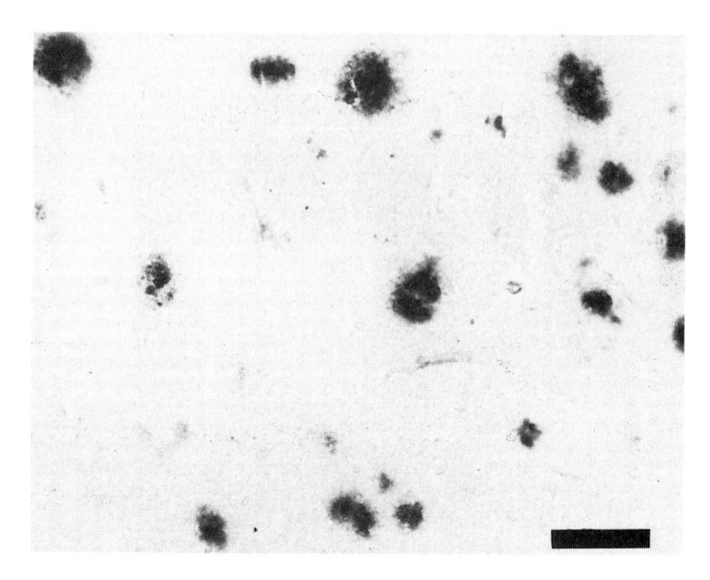

Figure 4. Immunocytochemical staining of cerebral cortex in a case of chronic renal dialysis showing the high density of immature senile plaques containing the A4 amyloid protein. Bar scale = 100 microns.

distribution pattern of aluminium consistent with its accumulation within pyramidal neurones in dialysis patients but not age-matched control cases (Fig 5). Intracellular accumulation of silicon was not observed. However, low level deposits of silicon were associated with the amyloid deposits in the immature senile plaques. The presence of immature senile plaques in only a third of patients with chronic renal failure and the absence of a correlation between gross brain levels of aluminium and the presence of plaques, suggest that there is not a simple relationship between exposure and the focal deposition of amyloid fibrils in plaque formation. Genetic or other factors may determine the susceptibility to such changes and as association between HLA-B40 antigen and vulnerability to dialysis encephalopathy has been reported.[45] Aluminium appears to accumulate mainly within neurones where it could influence either expression or processing of the A4 precursor protein, while silicon is present extracellularly associated with the A4 amyloid fibrils where it could prevent the removal of such deposits by glial or other cells.

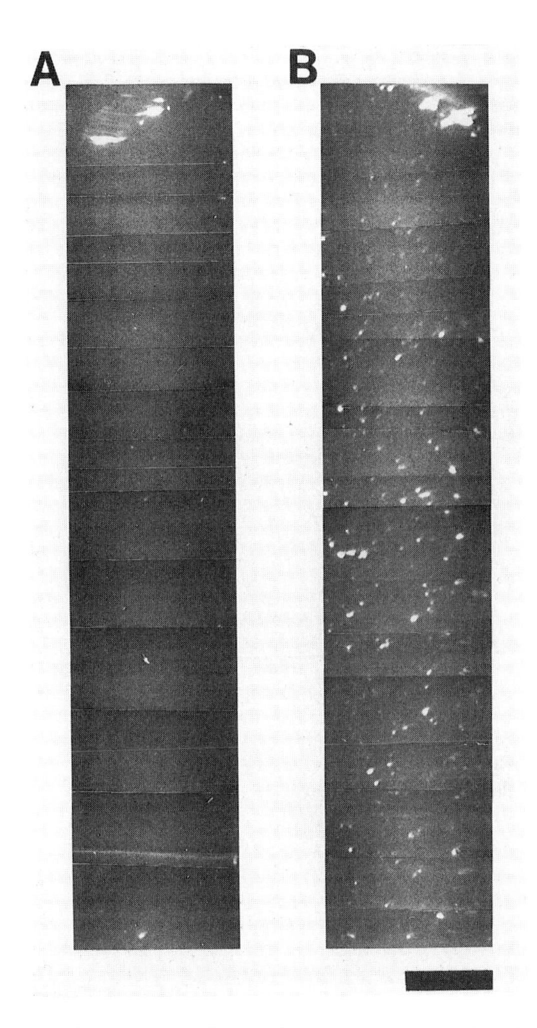

Figure 5. Imaging secondary ion mass spectrometry (SIMS) of the cerebral cortex in (A) control case (B) chronic renal dialysis, showing the laminar distribution of aluminium in what appear to be mainly pyramidal neurones and a few cells on the pial surface. Bar scale = 200 microns.

6 DISCUSSION AND CONCLUSIONS

Research on the cause of Alzheimer's disease has recently drawn attention to the molecular genetics of some familial forms. However, case-control studies and other epidemiological investigations have repeatedly shown that there is a high proportion of apparently sporadic cases. There is obviously a likelihood of underdetection of genetic factors in a disease with late onset but, in a recent study, cases with an onset at the age of 64 years or younger were found less likely to have a familial history of dementia than older cases.[46] Thus the existence of environmental risk factors in the aetiology of Alzheimer's disease cannot be excluded at the present time and there is more evidence for a possible association with aluminium than there is for any other agent. Four recent strands of evidence are consistent with the aluminium hypothesis, namely (i) the consistent and apparently specific deposition of this element in the form of aluminosilicate in the central region of the senile plaque core (ii) the likely mechanism of aluminium uptake into brain, involving transferrin mediated transport, which may explain the selective vulnerability of the brain regions affected in Alzheimer's disease (iii) the presence of high densities of immature plaques containing A4 amyloid in patients with chronic renal failure where there is a significantly increased brain content of both aluminium and silicon (iv) recent epidemiological studies,[47] discussed elsewhere in this volume, which suggest a geographical correlation between exposure to aluminium in water supplies and the risk of Alzheimer's disease. Although none of this evidence is proof of a causal relationship, the burden of evidence is suggestive and it is possible to propose mechanisms which could explain a pathogenic role for aluminium, particularly in relation to silicon deposition.

Imaging SIMS has confirmed the intracellular accumulation of aluminium in what appear to be pyramidal neurones, consistent with the view proposed by Birchall and Chappell in this volume and elsewhere[48] that there may be no efficient clearance mechanism and that critical levels may be reached in long-lived cells. Intracellular accumulation of aluminium is likely to result in a number of damaging consequences, including the possibility of effects on gene expression; actions on calcium dependent enzymes and calcium buffering systems; binding to phosphate groups on ATP, inositol triphosphate, phosphorylated cytoskeletal proteins etc. In particular, aluminium may influence the synthesis or post-ribosomal processing of those 'stress related' proteins which are produced in greater amounts in compromised cells, and which may include the A4 amyloid precursor protein.[49]

The death of a neurone loaded with aluminium may result in the relatively high local concentrations of free aluminium which could effect the polymerization of silicic acid, a process which occurs spontaneously when the concentration exceeds 100 parts per million but is greatly enhanced in the presence of aluminium.[50] It is possible that aluminium could substitute in such a silicon rich deposit with the eventual formation of an aluminosilicate which could serve as a nidus for the 'star-burst' accumulation of A4 amyloid fibrils which characterize the senile plaque core. Indeed it is possible that such amyloid deposits may reflect attempts by the brain to isolate aluminosilicate deposits which may be highly active catalysts of, for example, proteolytic activity.

Our findings do not support the view that aluminium is involved in the formation of neurofibrillary tangles, since such lesions were not present in cases of chronic renal dialysis, including some individuals with a marked elevation in brain content of aluminium, nor in preliminary studies have we been able to find evidence for intracellular accumulation of aluminium in tangle rich areas of Alzheimer brain.

The present studies do not resolve the issue of possible individual susceptibility to the neurodegenerative effects of aluminium and this problem may be elucidated when the affected gene in inherited Alzheimer's disease is characterized. If further epidemiological studies confirm an association between environmental exposure to aluminium and the risk of Alzheimer's disease it will be essential (i) to identify the genetic or other factors which explain why some individuals are vulnerable (ii) to quantify the risk associated with different sources and forms of aluminium (iii) to elucidate fully the mechanisms which underlie aluminium transport, accumulation and cellular damage in the nervous system, and to develop methods of preventing these processes.

REFERENCES

1. D.B. Calne, A. Eisen, E. McGeer and P. Spencer, Lancet, 1986, ii, 1067.
2. A.I.G. McLaughlin, G. Kazontis, E. King, D. Teare, R.J. Porter and R. Owen, Br. J. ind. Med., 1962, 19, 253.
3. A.C. Alfrey, G.R. LeGendre and W.D. Kaehny, New Engl. J. Med., 1976, 294, 184.
4. D. Perl, D. Gadjusek, R. Garruto, R.T. Yanagihara and C.J. Gibbs, Science, 1982, 217, 1053.
5. R. Garruto and Y. Yase, Trends Neurosci., 1986, 9, 368.
6. P.S. Spencer, P.B. Nunn, J. Hugon, A.C. Ludolph, S.M. Ross, D.N. Roy and R.C. Robertson, Science, 1987, 237, 517.
7. M. Kidd, Nature, 1963, 197, 192.
8. C.M. Wischik, M. Novak, H.C. Thogerson, P.C. Edwards, M.J. Runswick, R. Jakes, J.E. Walker, C. Milstein, M. Roth and A. Klug, Proc. Natl. Acad. Sci., 1988, 85, 4506.
9. J.T. Coyle, D.L. Price and M.R. Delong, Science, 1983, 219, 1184.
10. R.D. Terry and H.M. Wisniewski, 'Alzheimer's disease and related conditions', Eds. G. Wolstenholme and M. O'Connor, Churchill, London, 1970, p.145.
11. G.G. Glenner and C.W. Wong, Biochem. Biophys. Res. Comm., 1984, 122, 1131.
12. J. Kang, H.G. Lemaire, A. Unterbeck, J.M. Salbaum, C.L. Masters, K-H, Grzeschik, G. Multhaup, K. Beyreuther and B. Muller-Hill, Nature, 1987, 325, 723.
13. P. Ponte, P. Gonzalez-DeWhitt, J. Schilling, J. Miller, D. Hsu, B. Greenberg, K. Davis, W. Wallace, I. Lieburg, F. Fuller and B. Cordell, Nature, 1988, 331, 525.
14. R.E. Tanzi, A. McClatchey, E. Lamperti, L. Villa-Komaroff, J. Gusella and R. Neve, Nature, 1988, 331, 528.
15. N. Kitaguchi, Y. Takahashi, S. Shiojiri and H. Ito, Nature, , 331, 530.
16. I. Klatzo, H.M. Wisniewski and E. Streicher, J. Neuropath. exp. Neurol., 1965, 24, 187.
17. R.D. Terry and C. Pena, J. Neuropath. exp. Neurol., 1965, 24, 200.
18. J.C. Troncoso, L.A. Sternberger, N.A. Sternberger, P.N. Hoffman and D.L. Price, J. Neuropath. exp. Neurol., 1985, 44, 332.

19. N.A. Sternberger, L.A. Sternberger and J. Ulrich, Proc. Natl. Acad. Sci. USA, 1985, 82, 4274.
20. D.R. Crapper, S.S. Krishman and A.J. Dalton, Science, 1973, 180, 511.
21. D.R. Crapper, S.S. Krishman and S. Quittkat, Brain, 1976, 99, 67.
22. G.A. Trapp, G.D. Miner, R.L. Zimmerman, A.R. Mastri and L.L. Heston, Biol. Psychiat., 1978, 13, 709.
23. J.R. McDermott, A.I. Smith, K. Iqbal and H.M. Wisniewski, Neurology, 1979, 29, 809.
24. W.R. Markesbery, W.D. Ehman, T.I.M. Hossain, M. Alauddin and D.T. Goodin, Ann. Neurol., 1981, 10, 511.
25. B.E. Tomlinson, G. Blessed and M. Roth, J. Neurol. Sci., 1968, 7, 331.
26. N. Ward and J. Mason, J. Radioanal. Nuc. Chem., 1987, 113, 515.
27. D.P. Perl and A.R. Brody, Science, 1980, 208, 297.
28. S. Kobayashi, N. Hirota, K. Saito and M. Utsuyama, Acta neuropath., 1987, 74, 47.
29. J.M. Candy, A.E. Oakley, J. Atack, R.H. Perry and E.K. Perry, 'Regulation of transmitter function: basic and clinical aspects', Akademai Kiado, Budapest, 1984, p.183.
30. J.M. Candy, J.A. Edwardson, J. Klinowski, A.E. Oakley, E.K. Perry and R.H. Perry, 'Senile dementia of Alzheimer type', Springer, Heidelberg, 1985, p.183.
31. J.M. Candy, A.E. Oakley, J. Klinowski, T.A. Carpenter, R.H. Perry, J.R. Atack, E.K. Perry, G. Blessed, A. Fairbairn and J.A. Edwardson, Lancet, 1986, i, 354.
32. J.A. Edwardson, J. Klinowski, A.E. Oakley, R.H. Perry and J.M. Candy, Ciba Fdn. Symp. 121, Wiley, Chichester, 1986, p.160.
33. J.M. Candy, A.E. Oakley, F. Watt, G. Grime, J. Klinowski, R.H. Perry and J.A. Edwardson, 'Colloque Inserm-Eurage, Modern Trends in aging research', Libbey, London, 1986, Vol. 147, p.443.
34. J.M. Candy, A.E. Oakley, D. Gauvreau, P. Chalker, H. Bishop, D. Moon, G. Staines and J.A. Edwardson, Interdiscipl. Topics Geront., 1988, 25, 140.
35. R.H. Perry, A.E. Oakley, J.M. Candy, J.A. Edwardson and E.K. Perry, J. Neuropath. Appl. Neurobiol., 1985, 11, 500.
36. P.O. Ganrot, Environ. Health Perspect., 1986, 65, 363.

37. M. Cochran, J. Coates and S. Neoh, FEBS Lett., 1984, 176, 129.
38. R.B. Martin, J. Savory, S. Brown, R.L. Berthoff and M.R. Willis, Clin. Chem., 1987, 33, 405.
39. C.M. Morris, J.M. Candy, J.M. Court, C.A. Whitford and J.A. Edwardson, Biochem. Soc. Trans., 1987, 15, 498.
40. J.M. Candy, J.A. Edwardson, R. Faircloth, A.B. Keith, C.M. Morris and R.G.L. Pullen, J. Physiol., 1987, 391, 34.
41. R.G.L. Pullen, J.M. Candy, C.M. Morris, G.A. Taylor, A.B. Keith and J.A. Edwardson, J. Cerebral Blood Flow Metab., 1989, (submitted).
42. C.M. Morris, J.A. Court, A.I. Moshtaghie, A. Skillen, J.M. Candy, R.H. Perry and J.A. Edwardson, Biochem. Soc. Trans., 1987, 15, 891.
43. J.A. Edwardson, A.E. Oakley, G.A. Taylor, F.K. McArthur, J.E. Thompson, K. Beyreuther, C.L. Master and J.M. Candy, 'Proceedings of the International Symposium on Alzheimer's disease', University of Kuopio Publications, 1988, p.81.
44. B. Wolozin, A. Scicutella and P.D. Davies, Proc. Natl. Acad. Sci. USA, 1988, 88, 6202.
45. P. Garrett, S. Spencer, D. Mulcany, P. Hanly, J. O'Hare, J. Murnaghan, J. Hanson, J. Donohoe, M. Carmody and W. O'Dwyer, Proc. EDTA-ERA, 1985, 22, 360.
46. L. Shalat, B. Seltzer, C. Pidcock and E. Baker, Neurology, 1987, 37, 1630.
47. C.N. Martyn, D.J.P. Barker, C. Osmond, E.C. Harris, J.A. Edwardson and R.F. Lacey, Lancet 1989, i, 59.
48. J.D. Birchall and J.S. Chappell, Lancet, 1988, ii 1008.
49. J.M. Salbaum, H.-G. Lemaire, A. Weidmann, C.L. Masters and K. Beyreuther, EMBO J., 1988, 7, 2807.
50. R. Iler, 'The chemistry of silica. Solubility, polymerisation, colloid and surface properties', Wiley, New York, 1979, p.l.

An Epidemiological Approach to Aluminium and Alzheimer's Disease

C.N. Martyn

MRC ENVIRONMENTAL EPIDEMIOLOGY UNIT, SOUTHAMPTON GENERAL HOSPITAL, SOUTHAMPTON, SO9 4XY, UK

INTRODUCTION

Aluminium has no known biological function in higher animals and the discovery of localised high concentrations of the metal in two of the characteristic pathological features - senile plaques and neurofibrillary tangle bearing-neurones - of brains of patients with Alzheimer's disease suggests that it may be important in the causation of the disease[1,2]. Since it is impossible to administer aluminium containing compounds to humans under experimental conditions, this hypothesis can only be tested directly by epidemiological techniques. The epidemiological approach is based upon the idea that diseases do not occur at random but in patterns which reflect the operation of their underlying causes. The hypothesis that aluminium is involved in the aetiology of Alzheimer's disease leads to the prediction that rates of the disease will be higher in areas where the population is exposed to greater amounts of bioavailable aluminium than in areas where exposure is less.

Aluminium is found widely in the environment and it is estimated that, in Britain, between 5 and 10 mg are ingested each day[3]. However, only a very small proportion of this ingested aluminium is absorbed[4]. In many parts of Britain aluminium sulphate is used as a coagulant in the treatment of water in order to remove suspended particulate matter and coloured humic substances. Most of the added aluminium is removed in the process of clarification but residual amounts may pass into supply. Even in areas where residual concentrations are high, aluminium from drinking water forms only a small part of the total daily intake but, because the aluminium is largely uncomplexed, it may make a disproportionate contribution to the total amount absorbed from the gastrointestinal tract.

We have conducted a survey in 88 county districts within England and Wales with differing concentrations of aluminium in the water supply to examine the possible relation between exposure to aluminium from this source and the development of Alzheimer's disease.

METHODS

Details of residual aluminium concentrations in all important water sources supplying these districts over the previous decade were made available to us by the local water authorities. From these data mean residual aluminium concentrations for each county district were estimated.

The incidence of Alzheimer's disease, other causes of dementia and epilepsy in people aged 40 to 69 was estimated from the records of neuro-diagnostic centres serving these districts. Rates of disease for each diagnosis in each district were directly age standardised to the 40 to 69 year old population of England and Wales of 1983. The relation between age standardised rates of Alzheimer's disease, other categories of dementia and epilepsy, and mean concentration of aluminium in water was investigated using log linear and logistic regression analysis. Rates of disease in individual county districts were adjusted to compensate for differences in distance to the nearest neuro-diagnostic centre and for differences in the size of the population served by these centres.

RESULTS

We identified a total of 1985 patients with dementia, 445 of whom were classified as having Alzheimer's disease, and 2936 patients with epilepsy.

There was a positive relation between rates of Alzheimer's disease and residual water aluminium concentrations. Rates of Alzheimer's disease in districts where residual concentrations of aluminium exceeded 0.11 mg/l were 1.5 times higher than in districts where aluminium concentrations were less than 0.01 mg/l. This positive relation was not found for other causes of dementia or for epilepsy.

DISCUSSION

The results of this survey provide evidence of a relation between aluminium and Alzheimer's disease. A more complete account of this work, giving details of methods of estimating aluminium concentrations and rates of disease and a full description of the results, will be published elsewhere. However, it is worth pointing out here, that care is needed in the interpretation of our findings because, as in all epidemiological surveys, there is a possibility that the relation observed is due to the operation of some unknown confounding variable. Other studies in different populations are required for confirmation.

A further point which arises, concerns our ignorance of the way in which the human body absorbs and excretes aluminium. Future epidemiological studies should investigate some of the many other environmental sources of aluminium. If chemists can identify which of these are likely to have high bioavailability, such studies can be better targeted.

REFERENCES

1. J.M. Candy, A. E. Oakley, J. Klinowski et al. Aluminosilicates and senile plaque formation in Alzheimer's disease. Lancet, 1986, i: 354-357

2. D.P. Perl and A.R. Brody. Alzheimer's disease: x-ray spectrometric evidence of aluminium accumulation in neurofibrillary tangle-bearing neurones. Science 1980, 208: 297-299

3. Survey of aluminium, antimony, chromium, indium, nickel, thallium and tin in food. MAFF - the fifteenth report of the steering group on food surveillance. The working party on the monitoring of food stuffs for heavy metals. HMSO, 1985

4. P.O. Ganrot. Metabolism and possible health effects of aluminium. Environmental health perspectives 1986, 65: 363-441

The Chemistry of Aluminium and Silicon within the Biological Environment

J.D. Birchall and J.S. Chappell

IMPERIAL CHEMICAL INDUSTRIES PLC, PO BOX 11, THE HEATH, RUNCORN, CHESHIRE WA7 4QE, UK

1 INTRODUCTION

With oxygen, silicon and aluminium comprise the major elements of the earth's crust, mostly combined as the aluminosilicates of rocks, clays and soil minerals. Silicon and aluminium are both leached from these minerals with the leaching of aluminium being favoured by acidic conditions. The mean concentration of Al in rivers is 15μmol l^{-1} and that of Si 218μmol l^{-1} (1). The biosphere is thus exposed to both elements, and whilst Si is regarded by some authorities as an essential trace element - for review see reference (2), Al has been considered inessential and even harmless. However, the results of exposing patients on long-term haemodialysis to Al-rich dialysate (introducing Al directly into plasma) has shown Al to be responsible for the pathogenesis of a number of clinical disorders. The most significant of these are: (3) (4) (5)

(a) a disorder of bone - osteomalacic dialysis osteodystrophy

(b) a microcytic, hypochromic, non-iron deficient anaemia responding to low aluminium dialysate

(c) a progressive fatal neurological syndrome - dialysis encephalopathy or dialysis dementia.

The body burden of Al is in the range 30-60mg and normal plasma levels in the range 1.5 - 15μg l^{-1} although this can rise significantly in patients with chronic renal failure (ca 60 μg l^{-1}) and to higher levels still in patients on dialysis (>100 μg l^{-1}). Body burden may then reach several hundred mg(6). It is now standard practice to subject high Al water to ion exchange with reverse osmosis before it is used in dialysis. The implication of Al in these disorders is clear: their incidence is reduced by the removal of Al from water; the condition of severely affected patients is relieved by treatment with the aluminium chelator desferrioxamine[7] and aluminium is detected in the damaged organs for example, in osteomalacic bone at the cement line [4] [8].

The pathological changes produced in dialysis with high Al dialysate are the consequence of a massive assault on the system over a relatively short period. The discovery of aluminium localised in neurons bearing neurofibrillary tangles[9] and aluminosilicates at the core of senile plaques[10] [11] has prompted the questions: Is aluminium somehow implicated in the etiology of Alzheimer's disease and is A.D. a consequence of an insidious exposure to low levels of aluminium from water, food etc. over many years, perhaps allied to other intrinsic factors which determine absorption, transport etc? Whilst the detailed pathological features in the brain in dialysis encephalopathy and A.D. show differences - for example, the paired helical filament arrangement of neurofibrillary tangles in A.D. - "the rate of intoxication can modulate the expression and position of pathological change" [12].

The chemistry of aluminium is entirely relevant in that it will dictate the binding of aluminium to biological molecules and macromolecules and this will determine the absorption, transport and destination of aluminium and the molecular sites of biochemical lesions induced by the metal.

2 RELEVANT CHEMISTRY OF ALUMINIUM

The ionic radius of Al^{3+} is most like that of Fe^{3+} and its hydrolysis behaviour in aqueous solution is related to that shown by Fe^{3+}. The $[Al(H_2O)_6]^{3+}$ cation is stable only in acidic solution so that at pH values more basic than ca 3, hydrolysis occurs with the appearance of $[Al(H_2O)_5(OH)]^{2+}$ and $[Al(H_2O)_4(OH)_2]^{+}$. The formation of $Al(OH)_3$ dictates the limiting solubility of aluminium (at about pH 6.5) and at more basic pH values $Al(OH)_4^-$, the aluminate anion, is formed. At pH 7.4 the major species is considered to be $Al(OH)_4^-$. These species interact with various ligands - with aluminium preferring "hard" ligands (e.g. oxygen) - and those that need especially to be considered with respect to the in vivo context include:

- Fe^{3+}- binding proteins, especially transferrin (3g l^{-1} in plasma)

- Citrate (0.1 mmol l^{-1} in plasma)

- Phosphate (2 mmol l^{-1} in plasma: 10 mmol l^{-1} intracellular)

- Silicic acid (21μmol l^{-1} in plasma) [13]

It is important to consider silicic acid since it is a normal plasma component, hitherto neglected, and amorphous aluminosilicates have been found to be "a consistent feature of the senile plaque core"(14). At physiological pH silicic acid is uncharged (pK_1 >9) and unbound. The relative binding strength of the important ligands listed will determine the pathway of aluminium in vivo and its competition with other metal ions, e.g. Mg^{2+}. For example, aluminium inhibits hexokinase by forming ATP-Al^{3+} which competes with the active beta-gamma complex ATP-Mg^{2+}(15) (16). Although the Mg^{2+} concentration at 1-2 mmol l^{-1} is much greater than that of Al^{3+}, the latter binds to ATP 10^7 times more strongly than Mg^{2+} and thus can compete at very low concentrations. However, this competition must be modified in vivo by the presence of citrate, for the complex of Al^{3+} with citrate (log K = ca 18) is 10^5 times more stable than ATP-Al^{3+} and citrate activates the Al poisoned system in vitro (17). Thus the in vivo milieu will be very different from systems studied in isolation in vitro. Aluminium binds to the phosphate groups on DNA (18) and interacts with the phosphate groups of membrane phospholipids(19), these interactions depending on pH and the presence of other complexing molecules as illustrated for ATP.

The toxicity of aluminium to fish and to trees which is evident at the interface with the external environment (gill and root membranes respectively) may well result from binding of Al-cationic species to membranes. However, within the mammalian system the damage wrought by aluminium must involve strong binding to key sites in competition with other ions and ligands and a mechanism for amplification of damage from the foci. This is now considered.

3 THE PATHWAY OF ALUMINIUM

There is evidence that aluminium follows the Fe^{3+} pathway. It is bound in the iron transport protein transferrin(20); the microcytic hypochromic anaemia of dialysis suggests interference in iron metabolism; there is evidence of concomitant lysosomal storage of iron and aluminium and concomitant deposition(24). A simplified view of the iron pathway is:

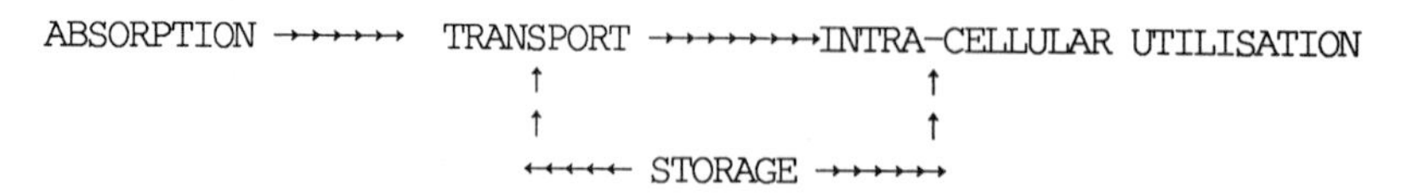

At pH 7.4 Fe^{3+} is highly insoluble (maximum free Fe^{3+} concentration 10^{-17}mol l^{-1}) and is solubilised and transported by the glycoprotein transferrin which has two M^{3+} binding sites per molecule of ca 77,000 Da. There is essential, concomitant binding of HCO_3^- and Fe^{3+} is very strongly bound (log K = ca 23). With 2- 3g l^{-1} plasma transferrin concentration and only one third binding sites filled there is significant binding capacity available for aluminium. Transferrin has been shown to bind aluminium with the need for concomitant HCO_3^- binding and it is significant that there are interactions in aqueous solution between HCO_3^- and aluminium (22) and aluminosilicate species (23). Transferrin binds 1μg per mg protein of both iron and aluminium. Aluminium is bound somewhat less strongly than iron but the binding is sufficiently strong for transferrin to dominate over all other ligands in extra-cellular fluid. Lactoferrin has also been found to bind aluminium (24) and the iron-binding proteins may be involved in the absorption of aluminium from the gut.

The function of transferrin is to transport iron to cells which have transferrin receptors on the plasma membrane, the receptor density varying significantly with cell type. There is some dispute as to events following transferrin binding to the cell surface receptor site but it is generally considered that transferrin is taken into the cell and its iron burden released in response to lower intra-cellular pH and/or reduction. There is considered to be an intra-cellular transit iron pool of low molecular weight iron chelators (e.g. citrate) within the cytosol from which iron is withdrawn for utilisation e.g. in mitochondria. Thus far, it is possible for aluminium to follow the same pathway and to be complexed within the labile iron pool. Its destiny thereafter is dictated by the subtle differences between iron and aluminium within the intra-cellular milieu.

4 THE INTRA-CELLULAR ENVIRONMENT

In the extra-cellular fluid, just as binding to transferrin prevents the interaction of Fe^{3+} with ligands such as OH^-, PO_4^{3-} etc. the related reactions of aluminium are suppressed by its binding in transferrin. (Were it not for this there would be extra-cellular deposits of iron III oxy-hydroxides and phosphates). Within the cellular environment, both iron and aluminium will be held in the labile pool from which iron is destined to be incorporated into haem and into enzymes that use iron with or without the haem binding configuration. Iron in excess is stored in the iron storage protein ferritin(25). This is a spherical shell of about 5 x 10^5 Da molecular weight comprising sub-units of ca 1.8 x 10^4 Da. There are channels within the shell for the entry and exit of iron. Each ferritin molecule can hold 4,500 atoms of iron as a "core" the approximate composition of which is $(FeOOH)_8(FeO{:}OPO_3H_2)$. The exact manner in which iron enters the protein shell of ferritin is much debated although *in vitro* studies suggest that it enters as Fe^{2+} and is oxidised within the shell and precipitated. It is also considered that the removal of iron from within the ferritin core involves reduction. If this is indeed the case, then ferritin would be unavailable as a sink for aluminium and although aluminium has been claimed to be present in ferritin isolated from aluminium-fed rats brains and from the brain ferritin of A.D. patients(26) the evidence is not totally convincing and chemical logic is somewhat against ferritin being an efficient sink for aluminium.

Intra-cellular aluminium, bound to citrate in the cytosol and unable to enter ferritin in competition with iron, will then become bound to binding sites that are stronger than the ligands holding aluminium (with iron) in the labile iron pool. Three need to be considered:

(a) binding in non-haem, iron-dependent enzymes

(b) phosphate; inorganic, phospholipids, phosphorylated proteins etc.

(c) silicic acid.

Non-haem Iron-Dependent Enzymes

Aluminium has been found to inhibit ferrioxidase (ceruloplasmin EC1.12.3.1.)(27) presumably competing for binding with Fe^{2+}. It has also been shown to reduce the activity of prolyl hydroxylase (EC1.14.11.2) in _in vitro_ experiments(28). The inhibition was greatest when the apo-enzyme was presented with aluminium before its essential co-factor iron. However, the competition is weak as shown by the absence of any inhibitory effect of aluminium in the presence of silicic acid which removes aluminium selectively as aluminosilicate. Except in the case of a massive assault by aluminium at high levels (or in isolated _in vitro_ experiments) it seems unlikely that aluminium could compete for binding in non-haem iron-dependent enzymes, especially in the presence of other strong chelators. Such a competition may be a factor in the microcytic hypochromic anaemia observed in dialysis, with high levels of aluminium interfering with a key step in haem synthesis - δ-amino levulinic dehydrogenase has been suggested as a step affected (5). Several non-haem, iron-dependent enzymes are involved in neurotransmitter synthesis but a more subtle mechanism must be invoked to account for neurotransmitter deficits than competitive inhibition of their synthesis by aluminium in the presence of more powerful chelators. As will be discussed, this probably involves aluminium binding to minority phosphate groups.

Binding by Special Phosphate Groups

The binding of aluminium by phosphate is inhibited by citrate (cf ATP) but there are molecules in which several phosphate groups exist in juxtaposition for co-operative binding. Phosvitin, for example, is an iron-binding phosphoglycoprotein of egg yolk rich in phosphorylserine residues. When saturated with iron, the Fe:P ratio is 1:2 which is attributed to the preferential binding of iron to adjacent phosphorylserine residues (c) rather than the alternative sites as illustrated:[29]

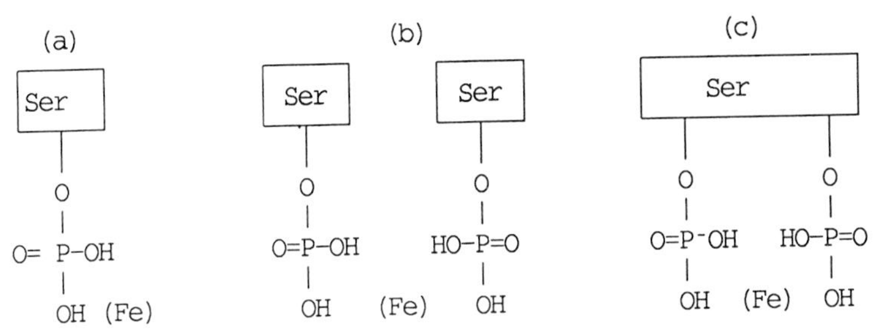

The binding of iron is strong (log K >17) and aluminium is also bound strongly[30]. In plants, phytic acid (inositol hexaphosphate) is a strong metal ion binder, occurring as the calcium/magnesium salt, phytin. Phytic acid is a probable sink for aluminium. When aluminium is present in equimolar (1 mmol l^{-1}) solutions of orthophosphate and citrate at pH 6.6, over 98% of the aluminium is bound to citrate whereas, when orthophosphate is replaced by inositol hexaphosphate less than 2% of aluminium is bound to citrate.

Of possible intra-cellular binding sites having phosphate groups geometrically favourable for strong binding the inositol phosphates are prime candidates. A computation of the binding of aluminium by ATP and by the adjacent (4,5) phosphates of inositol 1,4,5-triphosphate suggests a three-fold advantage of the binding energy for the inositol triphosphate. This finding may be of particular significance since the inositol phosphates are profoundly involved as messenger molecules in cellular signal transduction with a vital function in areas of the brain[31].

The Inositol Phosphate System

The reader is referred to authoratative reviews on this rapidly growing topic[32] [33]. Inositol lipids are located in the inner leaflet of the plasma membrane, phosphatidylinositol 4,5-biphosphate [PtIns(4,5)P_2] being one of these. In response to signals outside the cell, [PtIns(4,5)P_2] is cleaved to inositol 1,4,5-triphosphate [Ins P_3] and diacylglycerol as shown in Fig. 1. [InsP_3] stimulates Ca-mobilisation and almost certainly has other second messenger functions. It is converted to InsP_2 by inositol triphosphatase, terminating its messenger function. Diacylglycerol

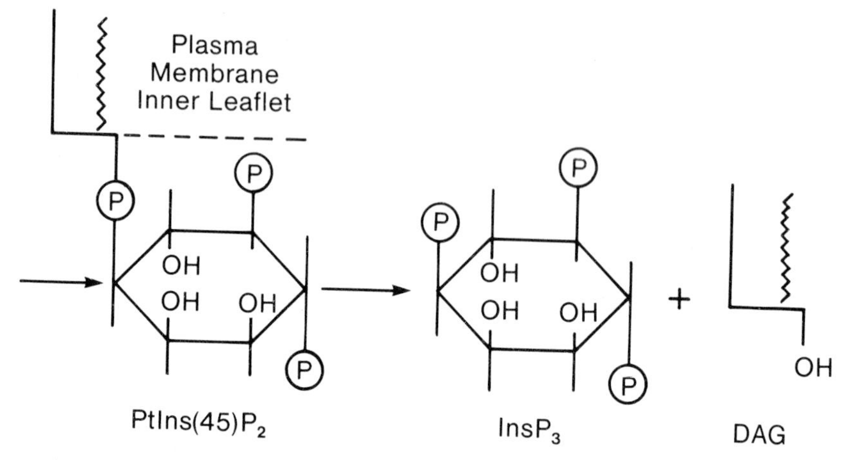

Figure 1.

activates protein kinase C, which itself has several messenger functions ([34]). There are enzymic pathways for resynthesising phosphatidylinositol from InP_2 and diacylglycerol to complete the cycle.

The strong binding of aluminium across vicinal phosphate groups seems likely to have a profound effect on the functions of the phosphatidylinositol system, disturbing Ca mobilisation, phosphorylation via protein kinase C and neurotransmitter release([31]). Such effects could provide the damage amplification required to account for profound changes wrought by very low concentrations of a toxic entity.

The prime characteristics of A.D. are: neurotransmitter deficit, neurofibrillary tangles as paired helical filaments within the neuronal cytosol and extra-cellular senile plaques with aluminosilicate at the core. The proteins of the tangles are abnormally phosphorylated and contain a fragment, tau, normally associated with microtubules([35]) ([36]). It may be significant that protein kinase C is reported to phosphorylate neurofilament proteins, adding to the evidence that this kinase is able to phosphorylate cytoskeletal elements([37]). Once proteins are phosphorylated, the binding of aluminium will reduce electrostatic repulsion and produce conformational change, in phosvitin, for example, toward the β-structure([38]).

If chemical rationale suggests that aluminium can disturb so fundamental and ubiquitous a system as the phosphatidylinositol-derived second messenger system, why is there not a general effect on the organism rather than a local effect in the brain? The plasma concentration of aluminium is low by comparison with iron, both being carried by transferrin. The flux of aluminium to cells will depend, not only on aluminium loading in transferrin, but also on transferrin receptor density. It may not be possible to deliver an effectively toxic dose of aluminium to a cell that has a short life (e.g. a reticulocyte) but possible, over years, to accumulate a toxic burden within a neuron. In the artificial conditions of dialysis, in which the flux of aluminium can be high, haem synthesis is affected by whatever mechanism, and neurological symptoms appear in months, perhaps an illustration of the effect of flux and timescale.

The Role of Silicic Acid, $Si(OH)_4$

Silicic acid is present in extra-cellular fluids (21μg l^{-1} in plasma) and within cells, being an uncharged small molecule free to diffuse. Silicic acid has a unique affinity for $Al(OH)_4^-$ which is structurally similar to $Si(OH)_4$ and the two interact to produce aluminosilicates. This is well-known in soil science in which dilute solutions containing aluminium and silicic acid (released by weathering from clay minerals) eventually precipitate aluminosilicate solids related to imogolite[39]. In dilute solution, the precipitation is slow and is retarded further by citrate and bicarbonate. However, it has been shown that, even in the presence of citrate, soluble aluminosilicate species exist with Si:Al ratio in the range 0.25-0.5[31]. In plasma, aluminosilicate species will not form as aluminium is bound to transferrin. When silicic acid, inorganic phosphate and aluminium are present together in dilute solution, there is a switch in the binding of aluminium at pH ca 6.6. Below pH ca 6.6 aluminium is bound to phosphate whereas above pH ca 6.6 - and reflecting the transition from cationic, octahedral aluminium to tetrahedral, anionic aluminium - aluminosilicate species are formed[31]. The transition is thus delicately balanced between the extra and intra-cellular environment. Within the cell, aluminium will predominantly be bound to phosphate and specifically to phosphate ligands more stable than citrate as discussed. In extra-cellular space aluminosilicate solids could form - in the absence of transferrin and especially at low citrate and bicarbonate levels - by the export of aluminium from dead neurons and its combination with silicic acid[23]. There is likely to be a residual affinity at the surface of such aluminosilicate deposits for interaction with phosphate groups on proteins which might thus be adsorbed locally leaving aluminosilicate at the core.

5 DISCUSSION

There is much speculation as to the role of aluminium in

Alzheimer's disease, in particular whether its presence within diseased neurons and, combined with silicon at the core of senile plaques, is a cause or consequence of disease. The presence of aluminium within the brain requires answers to questions as to its absorption, transport and binding. Can its binding to a particular site promote a biochemical lesion sufficiently critical for damage to be amplified? This paper has attempted to tackle such questions from the standpoint of the inorganic chemistry of aluminium and silicon.

The entry of aluminium into the system via the gastro-intestinal tract has been considered most often although the suggestion has been made, following experiments on rabbits, that the nasal-olfactory system may provide a mode of entry, with defects in the olfactory mucosa/olfactory bulb barriers leading to the influx of aluminium and silicon to parts of the brain(40). The parts of the brain affected by Alzheimer's disease - hippocampus, amygdala etc. have anatomical connections with the olfactory system. Aluminium intake via inhalation is said to be in the range 3-15µg d^{-1}. (In this context, it is interesting that a form of non-filarial elephantiasis in Ethiopia has been considered to be due to the entry of aluminosilicate particles into the system via skin lesions on the feet, their entry into the lymphatic system and a difference in the chemical manipulation of the inorganic particles by elephantiasic and normal subjects)(41).

Within plasma, aluminium transport seems akin to that of iron and hence its entry into cells (and across the blood-brain barrier) related to that of iron, with transferrin as the major carrier. Aluminium could co-exist with iron in an intra-cellular transit pool. However, aluminium cannot be utilised, for example, by mitochondria and ferritin will be largely unavailable as a sink. There may be no efficient clearance mechanism so that, slowly, the metal accumulates within cells although, normally, the flux is so low that a critical level is reached only in long-lived cells (the neuron). Other cell types are affected only when the flux is high, as in dialysis and systemic pathology results. Intra-cellular binding has then to be considered with sites binding more strongly than pool chelators being favoured. Attention is drawn to adjacent phosphate groups on inositol in which cation binding across vicinal phosphates is known to be preferred(42). The binding of aluminium across phosphate groups on the 4 and 5 positions and the slow ligand exchange characteristic of aluminium would then be expected to have strong effects on the manipulation of the inositol phosphate system and the second messengers derived from it, providing a mechanism for damage amplification. Aluminosilicate deposits at the core of plaques are seen, from the standpoint of the inorganic chemistry, as an end event with aluminium exported from dead neurons combining with silicic acid in the interstitial environment. What may determine individual vulnerability is beyond the scope of this paper, which seeks to review such chemical logic as may suggest key experiments.

REFERENCES

1. K.K. Turekian in Handbook of Geochemistry, Ed. K.H. Wedepohl, Springer-Verlag, Berlin, 1969, (1) 297-323.

2. J.D. Birchall in New Trends in Bio-Inoranic Chemistry, Ed. R.J.P. Williams and J.R.R.F. Da Silva, Academic Press, London 1978, (7) 209-252.

3. M.R. Wills and J. Savory, Lancet, 1983, ii, 29.

4. J. McClure and P.S. Smith, J. Pathol, 1984, 142, 293.

5. J.A. O'Hara and D.J. Murnagham, New Engl. J. Med., 1982, 306, 654.

6. K.C. Jones and B.G. Bennet, "Exposure commitment assessments of environmental pollutants" Monitoring and Assessment Research Centre report No. 3 Vol. 4, 1985, Kings College, University of London.

7. D.J. Brown, J.K. Dawborn, K.N. Ham and J.M. Xipell, Lancet, 1982, August 14, 343.

8. A.H. Verbueken, F.L. Van de Vyver, R.E. Van Grieken et al., Clin. Chem, 1984, 30,763.

9. D.P. Perl and A.R. Brody, Science,1980, 208,297.

10. J.M. Candy, A.E. Oakley, J. Klinoski et al,Lancet,1986, i, 354.

11. J. Edwardson, J. Klinowski, A.E. Oakley, R.H. Perry and J.M. Candy in Silicon Biochemistry, Ciba Foundation Symp. 121 John Wiley, Chichester, 1986, 160.

12. P.S. Spencer in "Selective Neuronal Death" Ciba Foundation Symp. 126. John Wiley, Chichester, 1987, 45.

13. J.W. Dobbie and M.J.B. Smith, Op.cit (ref 11), 194.

14. J. Edwardson et al, Op.cit (ref 11), 169.

15. F.C. Womack and S.P. Colowick, Proc. Nat. Acad. Sci. USA, 1979,76, 5080.

16. G.A. Trapp, Neurotoxicology, 1980, 1, 89.

17. R.E. Viola, J.F. Morrison, W.W. Cleland, Biochemistry, 1980, 19, 3131.

18. S.J. Karlik, G.L. Eichhorn, P.N. Lewis and D.R. Crapper, Biochemistry, 1980, 19, 5991.

19. R.H. Vierstra and A. Haug, Biochem. Biophys. Res. Commun., 1978, 84, 138.

20. G.A. Trapp, Life Sciences, 1983, 33, 311.

21. J. Bommer, R. Waldherr, P.H. Wieser and E. Ritz, Lancet, 1983, June 18, 1390.

22. Lars-Olof Öhman and W. Forsling, Acta Chem Scand, 1981, A35, 795.

23. J.S. Chappell and J.D. Birchall, Inorg. Chim. Acta., 1988, 153, 1.

24. A.A. Moshtaghie and A.W. Skillen, 617th Biochem. Soc. Meeting. Dundee 19-21 March 1987.

25. P.M. Harrison, G.A. Clegg and K. May in "Iron in Biochemistry and Medicine II" Ed. A. Jacobs and M. Worwood, Academic Press, London 1980, 4. 131.

26. J. Fleming and J.G. Joshi, Proc. Natl. Acad. Sci. USA, 1987, 84, 7866.

27. C.T. Huber and E. Frieden, J. Biochem., 1970, 245, 3979.

28. J.D. Birchall and A.W. Espie, Op.cit (ref 11)., 140.

29. J. Hegenauer, P. Saltman and G. Nace, Biochemistry, 1979, 18, No. 18, 3865.

30. J. Hegenauer, L. Pipley and G. Nace, Analytical Biochemistry, 1977, 78, 308.

31. J.D. Birchall and J.S. Chappell, Clin. Chem, 1988, 34, No. 2, 265.

32. M.J. Berridge and R.F. Irvine, Nature, 1984, 312, 315.

33. P.W. Majerus, T.M. Connolly, H. Deckmyn et al, Science, 1986, 234, 1519.

34. Y. Nishizuka, Science, 1986, 233, 305.

35. I. Grundke-Iqbal, K. Iqbal, Yunn-Chyn Tung et al, Proc. Natl. Acad. Sci. USA, 1986, 83, 4913

36. C.M. Wischik, R.A. Crowther, Brit. Med. Bull., 1986, 42, 51.

37. R.K. Sihag, Arco Y. Jeng and R.A. Nixon, FEBS Letters, 1988, 233, No. 1, 181.

38. G. Taborsky, J. Biol. Chem., 1968, 243, 6014.

39. V.C. Farmer, Op. cit. (ref. 11), 4.

40. D.P. Perl and P.F. Good, Lancet, 1987, May 2nd, 1028.

41. E.W. Price, and W.J. Henderson, Trans Roy Soc. Tropical Med. and Hygiene, 1978, 72, No. 2, 132.

42. H.S. Hendrickson and J.L. Reinertsen, Biochemistry, 1969, 8, 4855.

The Determination of Aluminium in Foods and Biological Materials

H.T. Delves, B. Suchak, and C.S. Fellows

TRACE ELEMENT UNIT, CLINICAL BIOCHEMISTRY, UNIVERSITY OF SOUTHAMPTON, SOUTHAMPTON SO9 4XY, UK

1 INTRODUCTION

Measurements of aluminium concentrations in body tissues and fluids are important for monitoring exposure levels for an increasing range of patients. The risk of toxicity from aluminium accumulation in patients with chronic renal failure on haemodialysis and receiving oral aluminium hydroxide therapy is well established. Currently attention is being focussed on the adventitious contamination of protein infusion fluids used in parenteral nutrition. The possible link between aluminium and Alzheimers disease has stimulated environmental studies at lower exposure levels. Additional background information on the general levels of dietary intakes of aluminium are provided by the analysis of a wide range of foods and beverages.

The role, in these studies, of atomic absorption spectrometry with electrothermal atomisation is described. Analyses of some specimens, eg. sera, require minimal preparation and small increases in concentration are easily detected. Some foods, particularly those with a high fat content eg. butter, present a greater analytical challenge. Nevertheless, with appropriate chemical preparation small differences in aluminium contents between salted and unsalted varieties can be determined. The high sensitivity of ETA-AAS gives a detection limit of $\sim$0.02 μg/g for most foods. This allows determinations of low levels occuring naturally, those introduced during processing as well as higher levels arising from the use of additives.

Measurement of Aluminium in Biological Materials

Routine analytical services for monitoring potential aluminium toxicity are provided by many hospital pathology laboratories. The patients most at risk are those who have chronical renal disease. As part of their treatment they receive oral doses of aluminium hydroxide gels to inhibit intestinal absorption of phosphate and they may also be exposed to aluminium from their dialysis fluids[1]. Aluminium toxicity has also occurred from the use of some infant formula feeds and intravenous fluids[2] and there are reports of very high concentrations of aluminium in some protein infusion fluids[3,4]. The pathological features of chronic aluminium toxicity include microcytic anaemia, bone disorders, dementia and even death[1]. The elevated concentrations of aluminium in serum associated with increased exposure (Figure 1) are poor predictors of these pathologies and provide only a limited, but nevertheless, useful guide in patient management. Other measures which have also proved useful but which have obvious limitations are the concentrations of aluminium in bone biopsies[5] and in cerebrospinal fluid[6].

Although many analytical techniques are available for measuring aluminium in body tissues and fluids most of those currently used in clinical laboratories are based on atomic spectroscopy. Emission spectroscopy using d.c. or r.f. plasmas as excitation sources and plasma source mass spectrometry have excellent sensitivities but are limited to a few specialised laboratories. The most commonly used methods employ electrothermal atomisation and atomic absorption spectrometry (ETA-AAS). The high analytical sensitivity of ETA-AAS, 16 pg Aℓ/0.0044A, enables the very low concentrations of aluminium normally present in human tissues (<1 μmol kg^{-1}) to be measured accurately with very small sample sizes eg. 1-5 μℓ serum. The main difficulties with clinical aluminium analyses by ETA-AAS are the effect of matrix composition on analyte sensitivity and contamination. The former is more easily controlled than the latter. The main interferences from biological samples are; volatilisation losses of $A\ell_2C\ell_2$ from chloride rich media, variable enhancement of analyte atomic signals from PO_4^{3-}, Ca^{2+}, and from carbonaceous residues formed during electrothermal decomposition. A variety of chemical reagents have been used to reduce as far as possible levels of chloride and of carbonaceous residue before atomisation of aluminium. Brown et al[7] precipitated serum proteins using HNO_3 yet still found that standard additions to different sera, especially uraemic sera, gave different calibration slopes. Others have used a variety of procedures to enable direct calibration with similarly treated aqueous standard solutions. Alderman and Gitelman[8] diluted sera, from 1+1 to 1+4, with a diluent containing $NH_4OH/Na_2H_2EDTA/H_2SO_4$/ Triton X100, ashed the sample in situ in argon at 1580°C and atomised at 2600°C using a molybdenum coated graphite tube. Others have used atomisation from a L'vov platform following matrix modification with $Mg(NO_3)_2$/Triton X100 [9] or $Mg(NO_3)_2$ with in situ oxygen ashing[10]. In my own laboratory in situ oxygen ashing has been found to be essential to achieve parallel slopes for different sera when $NH_4H_2PO_4$ is used as matrix modifier. This removes completely the carbonaceous residues from ETA-AAS analyses of sera, which would otherwise enhance in an uncontrolled way the rates of formation of aluminium atoms via reduction of the oxide. It does not however

Figure 1 Concentrations of Aluminium in Serum

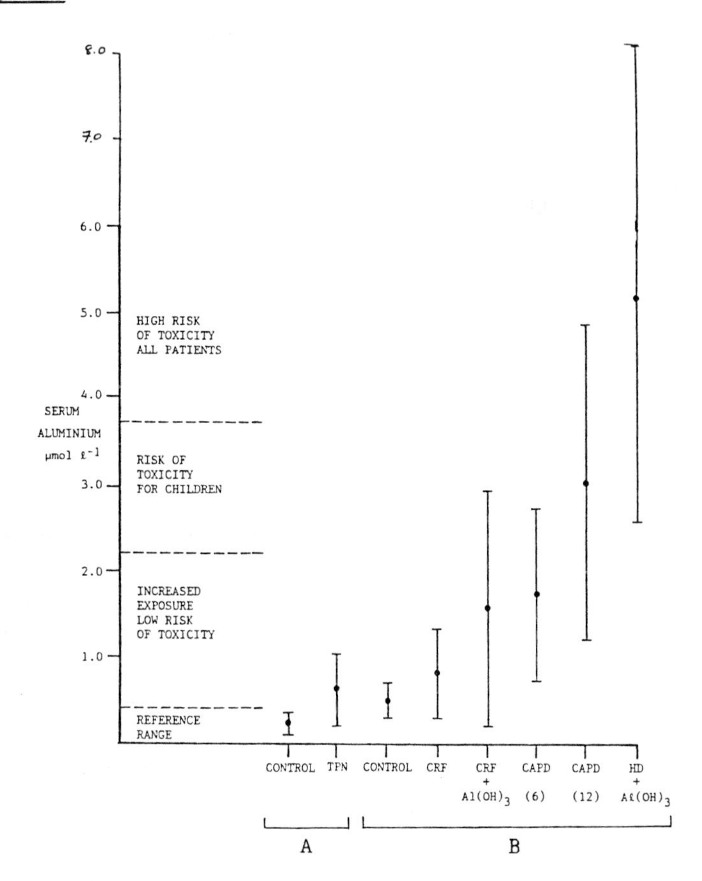

Data given as mean(•) ± 1SD (bars)

KEY

A	Unpublished observations authors laboratory
B	Data from Roberts et al[11]
CONTROL	Subjects with normal renal function not excessively exposed to aluminium
TPN	Patients on total parenteral nutrition
CRF	Patients with chronic renal failure
+$A\ell(OH)_3$	Patients receiving oral aluminium hydroxide
CAPD(6) CAPD(12)	Patients converted to continuous ambulatory peritoneal dialysis after six months(6) or twelve months(12) haemodialysis therapy
HD	Patients on haemodialysis

Figure 2 Aluminium in Serum ETA-AAS

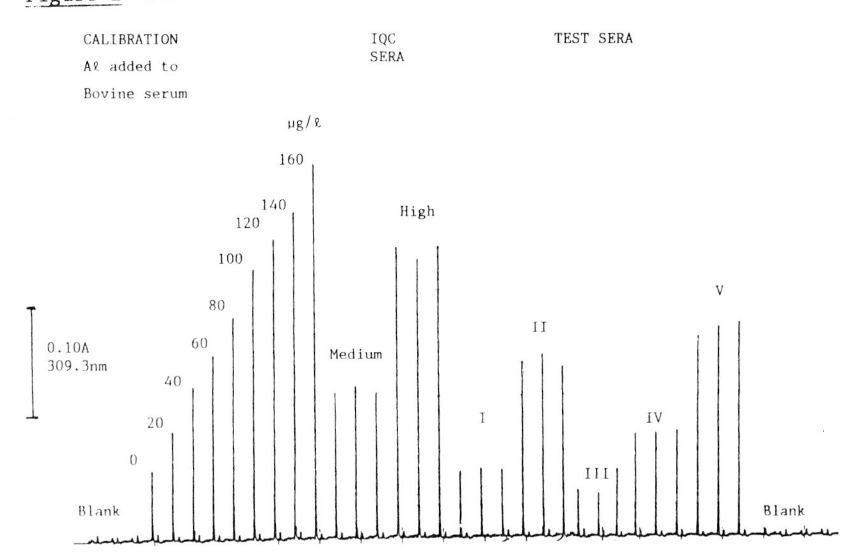

Chart recording of background corrected atomic absorption signals from aluminium in sera. Samples were prepared by diluting 100 μℓ sera, 1+1+2 with 1% m/v $NH_4H_2PO_4$ and H_2O. Calibration was effected by using similarly diluted bovine serum except 200 μℓ H_2O was replaced with 100 μℓ aqueous Aℓ standard plus 100 μℓ H_2O. Electrothermal atomisation conditions for 10 μℓ injection volumes are given below.

STEP	T/°C	RAMP/s	HOLD/s	GAS	GAS FLOW/ml min^{-1}
DRY	140	1	30	Ar	300
	160	50	1	Ar	300
ASH	550	10	30	O_2	50
DESORB	950	10	20	Ar	300
ASH	1200	5	20	Ar	300
COOL	20	1	30	Ar	300
ATOMISE	2650	0	5	Ar	10
CLEAN	2700	1	3	Ar	300
COOL	20	1	20	Ar	300

compensate for the effect of matrix on analyte sensitivity and matrix matched standards are needed. Figure 2 shows the atomic signals from part of an analytical run for serum aluminium measurements. It is essential to eliminate all sources of contamination, maintain a high standard of cleanliness and pay strict attention to details because of the ubiquity of aluminium. A single dust particle of 50-100μ diameter can contain 0.2-2 ng Aℓ and on falling into a 1ml volume of solution prepared for analysis would produce signals greater than those seen for the blanks in Figure 2. It has been known for a days work to be ruined by contamination caused by tearing a paper towel in the vicinity of uncapped solutions. Analyses are usually done in triplicate and a good within-sample relative standard deviation (RSD) see Figure 2 taken as freedom from random contamination. The analyses are controlled by accepting as valid only those results bounded by valid data for concurrently analysed internal quality control (IQC) specimens. Application of this IQC protocol ensures accuracy of the analytical data which is monitored by participation in national external quality assessment programmes. The detection limit is 0.04 μmol ℓ^{-1} (1 μg/ℓ) and the between batch SD is 0.079 at 0.56 μmol ℓ^{-1} and 0.045 at 2.93 μmol ℓ^{-1}.

Additional information on aluminium exposure has been provided by analyses of bone[12], urine[13]. Bone samples are usually extracted with petroleum ether to remove fatty material before digestion with HNO_3. Monitoring dialysis fluids to prevent excessive exposure to aluminium is becoming increasingly important. Halls and Fell[14] successfully analysed a range of dialysates by ETA-AAS using addition of 2% $^v/_v$ HNO_3 as modifier to prevent chloride mediated losses of aluminium during electrothermal decomposition. Although most dialysates contained <0.04 μmol ℓ^{-1} (1 μg/ℓ) some contained up to 31 μg/ℓ. Analysis of dialysis concentrates which contain ∿40% $^m/_v$ dissolved solids cannot be done by direct dilution without losing analytical sensitivity. However a simple chelating resin ion exchange separation[15] allows analyses down to 0.08 μmol ℓ^{-1} (2 μg/ℓ).

Measurement of Aluminium in Foods

Concern over the possible relation between environmental aluminium exposure and Alzheimers disease[16] has prompted studies of all forms of input of this element including that from foods. Some foods such as tea accumulate naturally high levels of aluminium. Others contain aluminium as a result of adventitious contamination during processing and packaging and from the use of permitted additives such as aluminium phosphates, aluminosilicates and aluminium colorants[17,18]. The concentration of aluminium in food as purchased ranges from <0.7 nmol kg^{-1} (<0.02 μg g^{-1}) up to ∿600 mmol Kg^{-1} (1.6% $^m/_v$). The latter figure being the addition of aluminosilicates to some icing sugars. As part of a study of dietary intakes of aluminium in the United Kingdom our laboratory is measuring total concentrations of this element in a wide range of foods, beverages and in total diet samples. The analytical challenge varies with the nature of the food matrix and its aluminium content. Examples discussed here will be of three types of food or beverage: (a) liquid samples analysed by direct dilution and ETA-AAS; (b) some liquid/solid samples requiring minimal chemical treatment eg. acid digestion before analysis; (c) high fat content foods requiring additional treatment.

(a) Direct Analysis. Milk, fruit juices/drinks and infusions of tea and coffee can easily be analysed for aluminium by ETA-AAS following a simple dilution with 1% $^{m}/_{v}$ $(NH_4)_2HPO_4$ and using ETA conditions similar to those given in the footnote to Figure 2. For most samples a 1+3 dilution is used but some of the very high concentrations found in fruit juices and in tea require up to 100 fold dilutions. The atomic absorption signals as a function of time after start of the atomisation stage (Figure 3a) show very little differences between the diluted beverages and the aqueous aluminium standards. These data indicate that matrix interferences are negligible and that calibration can be done using simple aqueous standards. However, for the current investigations of aluminium in foods all analyses in our laboratory are done by standard additions as well as by comparison with matrix-matched or simple aqueous standards. This protocol affords additional confidence in analytical data.

The range of aluminium concentrations found in the liquid samples analysed thus far can be seen from Figure 4. The aluminium content of milk samples ranged from 0.4-1.2 μmol ℓ^{-1} (11-35 μg/ℓ) whereas concentrations in fruit juices and drinks were all higher and reached 38 μmol ℓ^{-1} (1029 μg/ℓ). There was no evidence that these higher concentrations originated from aluminised cartons.

(b) Samples Requiring Acid Digestion. Many foods, eg. carrots, potatoes, apples, flours, tea leaves, coffee powder, can be oxidised with concentrated nitric acid to yield clear aqueous solutions ideally suited for analysis by ETA-AAS. Occasionally a slightly cloudy solution is obtained which is probably SiO_2 from the quartz conical flasks used since recoveries of added aluminium are quantitative. The atomic absorption profiles for solutions of digested foods diluted with 1% $NH_4H_2HPO_4$ are similar to that for a similarly diluted aqueous standard (Figure 3b). The small differences indicate some slight matrix interferences but none which would cause substantial analytical bias. The calibration graphs for carrots and flours (Figure 5) confirm the absence of any matrix interferences on analyte sensitivity. It should be noted that all of the atomic signals shown in Figures 3a, 3b and 3c occur within the time window of 1.0-2.5s after start of atomisation stage. This time period is close to the 1.3-3.0s period cited by Falk[19] as that time during which the variation temperature along the length of the graphite tube furnace is a minimum and during which the concentration of

Figure 3 Comparison of Aluminium Atomic Absorption Signals from Different Types of Foods and Beverages

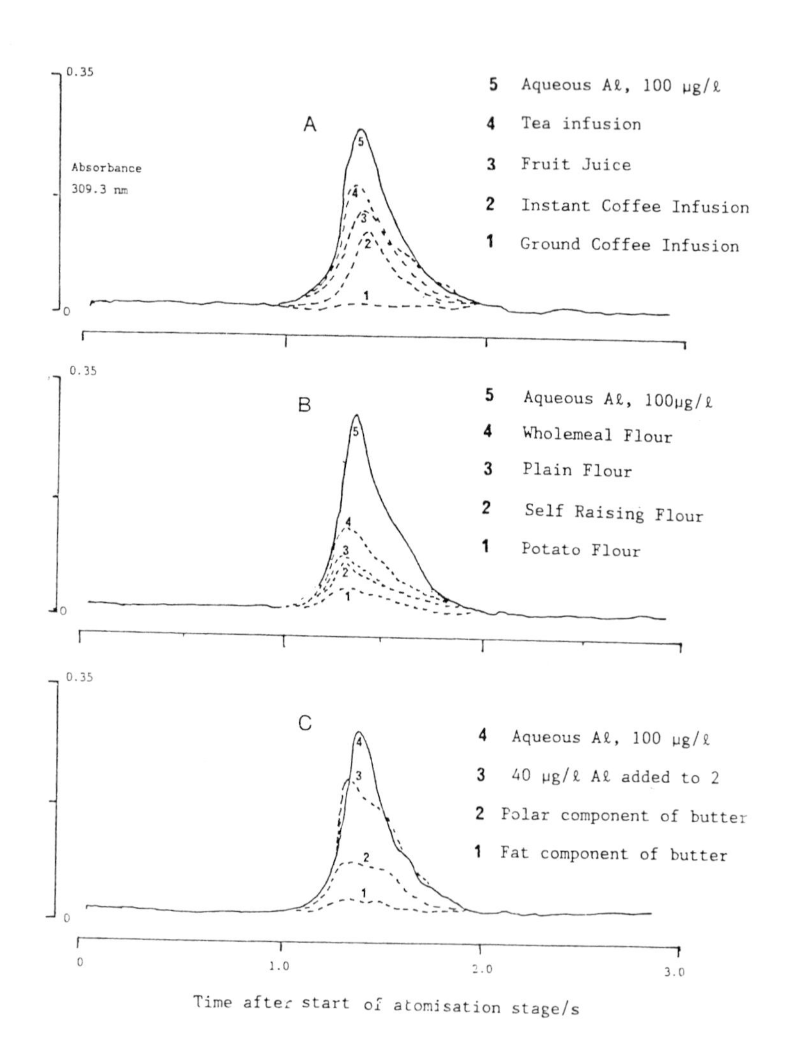

Figure 4 Aluminium Concentrations in Milk, Fruit Juices and Fruit Juice Drinks

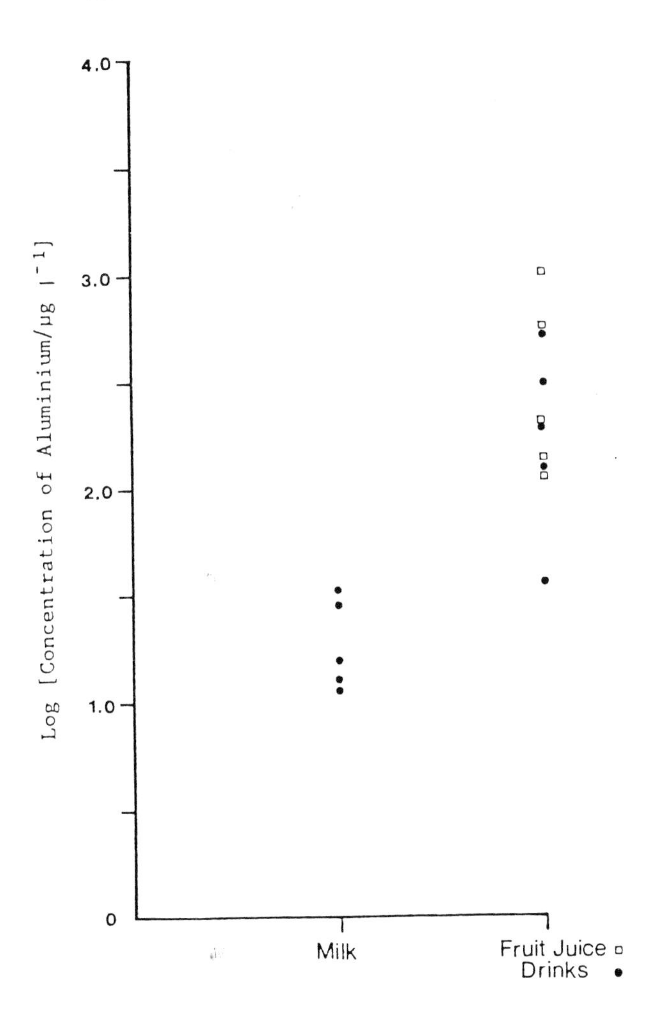

The observed concentrations of aluminium were: milk sample, 11-35 μg/ℓ; fruit drinks, 37-527 μg/ℓ and fruit juices, 116-1029 μg/ℓ.

gaseous carbon compounds is also minimum[20]. These conditions give maximum control of the physical and chemical reactions that would alter the free atom concentrations within the furnace and are thus optimum for absorption measurements.

(c) High Fat Content Foods. Acid digestion of foods such as butter, margarine, cheese using nitric acid mixtures is potentially hazardous because of their high fat contents. Although H_2SO_4/H_2O_2 mixtures have been used to oxidise these types of sample the reactions on additions of 50% H_2O_2 to hot biphasic fat/sulphuric acid mixtures are somewhat vigorous and possible loss of sample is only one disadvantage. An alternative approach to preparation of this type of sample is to remove the fat by solvent extraction with petroleum ether and back extract into 0.1M HNO_3 any fat-soluble aluminium. The polar, ether insoluble residue is easily and safely oxidised with concentrated nitric acid. Measurement of aluminium in both phases, using ETA-AAS, shows most of the metal to be with the polar constituents (Figure 3c). The absence of matrix interferences with analysis of butter and of margarine is seen from the calibration graphs (Figure 6a), but there is a small suppresive interference with analysis of cheese (Figure 6b). The precision of this method of preparation and analysis is shown by the ability to detect small changes in aluminium content of salted and unsalted butter, and of the inner and outer layers of cheese (Table 1). The latter is due to the aluminium lake in the wax coating around Edam cheese. Higher concentrations of aluminium have also been found in the skin of apples (4.2 $\mu g\ g^{-1}$) compared with the flesh (0.09 $\mu g\ g^{-1}$).

Table 1 Detection of Small Changes in Aluminium Content of Cheese and Butter

Sample	Variable	Concentration of Aluminium, $\mu g\ g^{-1}$ mean ± SD (N)
Butter	Unsalted	<0.05 ± * (5)
	Salted	1.23 ± 0.06 (6)
Edam Cheese	Inner Core	0.15 ± 0.03 (3)
	Outer 5mm	0.34 ± 0.02 (3)

* All data below detection limit

Accuracy of Analyses of Aluminium in Foods. The quantitative recovery of aluminium added to a wide range of foods (Table 2) indicates the degree of accuracy of the analytical procedures outlined here. Of particular note is the good recovery of low concentrations, of aluminium added to butter and subjected to two solvent extraction stages and a nitric acid oxidation.

Figure 5 Calibration Curves for Easily Oxidised Food Samples

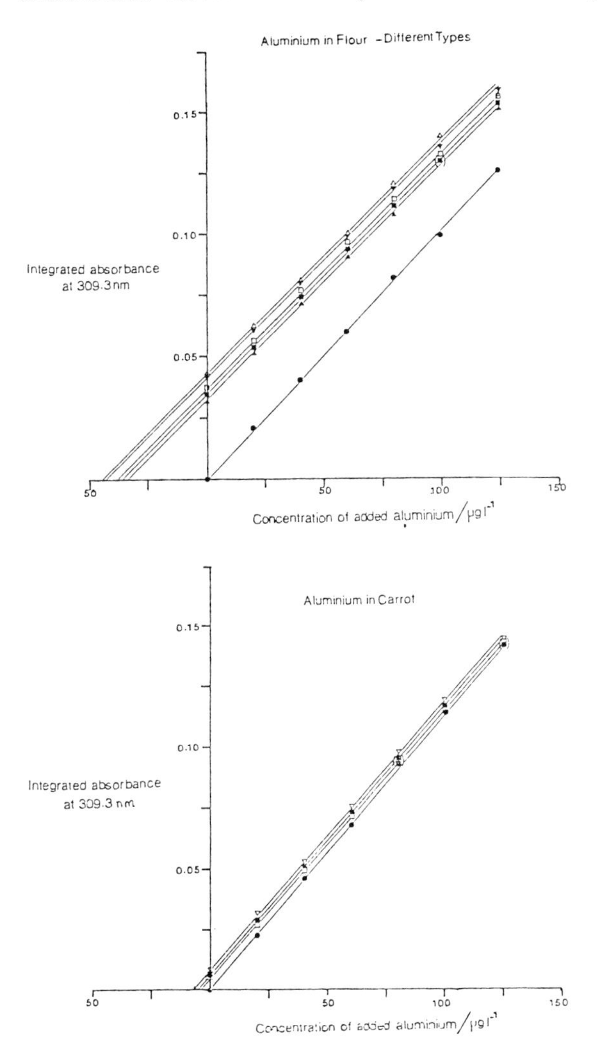

Above data show parallel nature of calibration curves obtained using standard additions to HNO_3 oxidised solutions of 5 different types of flour (A) and 3 different types of carrot (B). In each figure the lowest curve for analysis of aqueous aluminium standards has a similar slope to that obtained in presence of the oxidised food matrix.

Figure 6 Calibration Curves for High Fat Content Foods

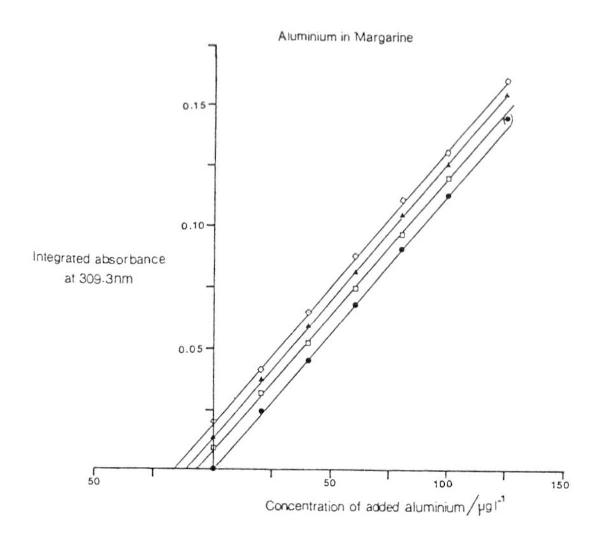

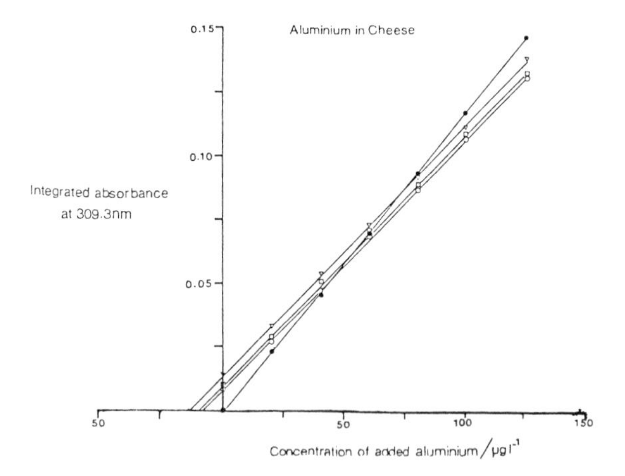

The upper Figure (A) shows the parallel nature of calibration curves obtained in presence of oxidised polar constituents of 3 different margarines and the lowest curve for aqueous standards. The fatty matrix in each case contained no aluminium and gave a response coincident with the lowest curve. Similar data are obtained for butter samples.

The lower Figure (B) shows suppressive effect of polar constituents of 3 different cheeses on calibration curve even after oxidation and matrix modification. Again lowest curve is for aqueous standards.

Table 2 Recovery of Aluminium Added to Foods and Beverages

Sample	Per-cent of added aluminium recovered Mean ± SD (N)	Concentration range of added Al $\mu g\ g^{-1}$
Milk	97.8 ± 2.9 (5)	0.02 - 0.10
Fruit Juice	100.8 ± 4.2 (5)	0.05 - 0.15
Coffee (Instant)	100.0 ± 4.9 (5)	0.40 - 2.0
Potato	99.7 ± 5.9 (3)	0.40 - 1.6
Rice Flour	100.1 ± 5.2 (3)	0.8 - 1.6
Butter	99.0 ± 2.7 (4)	0.8 - 3.2
Margarine	102.9 ± 2.7 (4)	0.8 - 3.2
Cheese	99.4 ± 2.1 (4)	0.8 - 3.2

Another indication of analytical accuracy is the good agreement between results obtained by two different calibration procedures (1) standard additions and (2) direct measurement against aqueous/matrix matched standards (Figure 7). The observed concentrations, for 68 different samples of food, range from 0.02 $\mu g\ g^{-1}$ up to 1500 $\mu g\ g^{-1}$. A comparison of aluminium analyses done using the methods described here and by those in other workers laboratories is shown in Table 3 for NBS mixed diet reference material and for two infant formulae milk powders. In all cases the agreement is good, although data from our laboratory appear to have a small negative bias.

Table 3 Concentration of aluminium, $\mu g\ g^{-1}$

	Southampton	Other laboratory
NBS, Mixed diet No. 8431	3.9 ± 0.2 (2)	4.4 ± 1.1 (Mean and certified range)
Infant formula 1	4.5 ± 0.5 (3)	4.9 ± 0 (2) (Massey[21])
Infant formula 2	0.61 ± 0.06 (3)	0.8 ± 0.1(2)

Significance of Aluminium in Food. The data in Figure 7 show the very low (<0.02 $\mu g\ g^{-1}$) endogenous levels of aluminium in some foods such as eggs and the high levels naturally present in tea (∿100 $\mu g\ g^{-1}$). It is also interesting to note the increase in aluminium content with food processing/packaging e.g. instant coffee has more aluminium than ground coffee (2.0 vs 0.4 $\mu g\ g^{-1}$). The use of food additives has produced high concentrations (50-100 $\mu g\ g^{-1}$) in some custard powders and up to 1500 $\mu g\ g^{-1}$ in some flour mixes. Of

Figure 7 Comparison of Concentrations of Aluminium Obtained for Different Food Samples by the Methods of Standard Additional and Direct Calibration

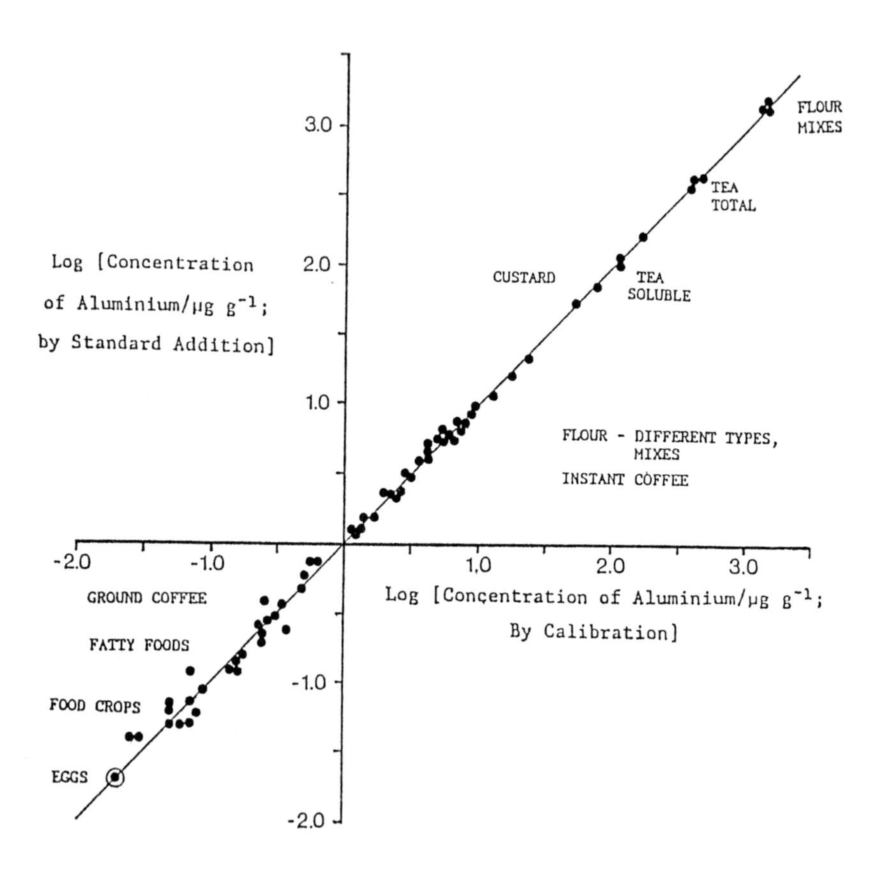

the 68 different samples of food/drinks, 29(43%) contained less than 1 μg g^{-1}Aℓ, 56(82%) less than 10 μg g^{-1} and 61(90%) less than 100 μg g^{-1}. These data are simply total concentrations and do not give any indication of the uptake of aluminium following ingestion of these foods. One would expect a much greater degree of intestinal absorption of aluminium from ingested fruit juices containing citrate than from ingestion of foods containing aluminosilicates or aluminium phosphates. The relative bioavailability of aluminium from different foods needs to be established. This could easily be assessed by measuring the increase in plasma aluminium concentrations following oral intakes. A similar approach but using increases in urinary excretion of aluminium has been used to assess uptake of aluminium from tea[22].

2 CONCLUSIONS

Although aluminium toxicity is now recognised as an iatrogenic complication in the treatment of chronic renal disease there remains the need for a reliable diagnostic indicator of potential toxicity. Total serum concentrations give useful but limited information. It is probable that measurements of the different aluminium carrier species in blood serum, particularly those of low molecular weight (<5-10 kDa) will be of greater clinical value. An analogous situation exists for interpreting the significance of aluminium in foods. Total concentrations give limited information and measures are needed of that fraction which is capable of being absorbed in the gastro-intestinal tract.

Accurate analytical methods for measuring aluminium in biological samples and in foods have been available since the mid 1970's following the developments in ETA-AAS by L'vov[23] and Massman[24]. At present the most pressing analytical requirements are for a wide range of certified reference materials for foods and biological specimens. These should have well established and realistic concentrations of aluminium. By using these materials for internal quality control of analyses obtained with modern atomic spectroscopic instrumentation, it will be possible to guarantee the analytical accuracy of data obtained in speciation and bioavailability studies. These data would lead to a greater understanding of the significance of aluminium in foods and of the role of this element in human disease.

Acknowledgement

We are grateful to Ministry of Agriculture Fisheries and Food for financial support for some of the work presented here.

References

1. DNS Kerr and MK Ward in 'Aluminium and other trace elements in renal disease' ed. A Taylor publ. Bailliere Tindall London, 1986, p.1.

2. M. Freundlich, G. Zillereulo, C. Abitbol and J. Strauss, Lancet, 1985, ii, p.527.

3. E.R. Maher, E.A. Brown, J.R. Curtis et al, Br.Med.J., 1986, 292, 306.

4. D. Maharaj, G.S. Fell, D.F. Boyce et al, Br.Med.J., 1987, 205, 693.

5. A.C. Alfrey, G.R. LeGendre and W.D. Kaehny, N.Eng.J.Med., 1976, 294, 184.

6. D. Brancaccio, O. Bugrami, L.Pacini et al in 'Aluminium and other trace elements in renal disease' ed. A. Taylor, publ. Bailliere Tindall, London, 1986, p.19.

7. S. Brown, R.L. Bertholf, M.R. Wills and J. Savory, Clin.Chem., 1984, 30, 1216.

8. F.R. Alderman and H.J. Gitelman, Clin.Chem., 1980, 26, 258.

9. F.Y. Leung, C. Bradley, W. Slavin and A.R. Henderson in 'Aluminium and other trace elements in renal disease' ed A. Taylor publ. Bailliere Tindall, London, 1986, p.296.

10. M. Bettinelli, U. Baroni, F. Fontana and P. Poisetti, Analyst, 1985, 110, 19.

11. N.B. Roberts, D. Fairclough, S. McLaughlin and W.H. Taylor. Ann. Clin.Biochem., 1985, 22, 533.

12. J. Smeyers-Verbeke and D. Verbeelen, Clin.Chem., 1985, 31, 1172.

13. F.Y. Leung and A.R. Henderson, Clin.Chem., 1982, 28, 2139.

14. D.J. Halls and G.S. Fell, Analyst, 1985, 110, 243.

15. C.S. Fellows and H.T. Delves, 'Unpublished observations'.

16. C. Martyn. Paper presented at Royal Society of Chemistry Meeting on Aluminium in Food and in the Environment, May 1988.

17. S. Fairweather-Tait, R.M. Faulks, S.J.A. Fatemi and G.R. Moore, Human Nutrition: Food Sci. and Nutr., 1987, 41F, 183.

18. J.E.A.T. Pennington. Food Additives and Contam. 1988, 5, 161.

19. H. Falk, A. Glismann, L. Bergann et al, Spectrochim Acta Part B, 1985, 40, 533.

20. B.V. L'vov, J.Analyt. Atom Spectrom., 1987, 2, 95.

21. R. Massey, (MAFF) Personal communication to HTD, 1987.

22. K.R. Koch, M.A.B. Pougnet and S. de Villiers, Nature, 1988, 333, 122.

23. B.V. L'vov, Spectrochim.Acta, 1961, 17, 761.

24. H. Massmann, Spectrochim.Acta Part B, 1986, 23, 215.

Aluminium in Foods and the Diet

J.C. Sherlock

MINISTRY OF AGRICULTURE, FISHERIES AND FOOD, GREAT WESTMINSTER HOUSE, HORSEFERRY ROAD, LONDON SWIP 2AE, UK

The past few years have seen an upsurge of interest in aluminium and human health. It is established that aluminium is responsible for 'dialysis dementia' in patients undergoing dialysis for kidney disease. In these cases aluminium enters the blood stream directly from the water used for dialysis. Patients suffering from 'dialysis dementia' have elevated concentrations of aluminium in brain tissue. Similarly, patients suffering from Alzheimer's disease have elevated concentrations of aluminium in some neurones. Alzheimer's disease and 'dialysis dementia' have some symptoms in common and therefore it is suspected that aluminium may be a cause of Alzheimer's disease. It is estimated that as many as several hundred thousand elderly people in the UK suffer from Alzheimer's disease. Because of the association between aluminium and Alzheimer's disease there is a need for information about the sources of exposure of man to aluminium: these sources inevitably include food and drink. In addition to this it is thought[1] that bottle-fed infants may be exposed to much greater intakes of aluminium than are breast-fed infants. This could be of concern since infants will tend to have a higher permeability of the blood-brain barrier to aluminium, a higher uptake of aluminium via the still-developing gastro-intestinal tract, and poorer excretion of aluminium through immature kidneys[2].

The need for information about exposure to aluminium in food and drink, and with particular reference to infant formulae is what prompted production of this paper. The following paragraphs present a summary of information on aluminium in foods and the diet and discuss factors which may influence aluminium intake.

1 ALUMINIUM IN FOODS

Table 1 presents information on aluminium in vegetables and cereals. The information is not all that is available but is a representative selection. All the data have been rounded to the nearest digit. The most recent data present the lowest concentrations; this is almost certainly a reflection of improving analytical methodology. The indications are therefore that vegetables and cereals contain aluminium at concentrations in the range 0.1 to 10 mg/kg with a strong suspicion that the 'true' values lie at the lower end of the range. The data also indicate that soil contamination might be responsible for some of the elevated concentrations of aluminium found in vegetables. Taking a typical soil aluminium concentration as 50,000 mg/kg (that is 5%), then as little as 1 g of soil in 1 kg of plant material would increase the apparent concentration of aluminium in the plant to 50 mg/kg.

Table 2 presents information on aluminium in meat, dairy products, fruit and tea infusions. Once again the general picture which emerges is one of aluminium concentrations in the range 0.1 to 10 mg/kg,with the majority of the concentrations at the lower end of this range. It is reasonable to suspect that the differences in the data for specific foods presented in Tables 1 and 2 are influenced by the limit of detection of the method of analysis rather than true differences in concentrations. The more sensitive methods of analysis used by Delves(8) and Sullivan(7) tend to yield lower concentrations. It seems safe to say that tea infusions are likely to contain 1-5 mg/l of aluminium and may, therefore, be an important source of dietary aluminium.

Table 1 Aluminium in vegetables and cereals (all concentrations in mg/kg fresh weight)

	Reference source					
Food	Koivistoinen (3) 1980	Teraoka (4) 1981	MAFF (5) 1985	Greger (6) 1985	Sullivan (7) 1987	Delves (8) 1988
Cabbage	<1	2	2	0.1	0.1	-
Carrots	3	5	26*	-	-	0.05
Celery	1	1	6	-	-	-
Lettuce	3	2	90*	1	-	-
Onion	2	1	2	-	-	-
Potato	2	1	8	2	0.2	<0.2
Wheat	7	7	2	-	-	-
Rice	62	1	-	2	0.2xx	-

*Possible soil contamination xxSoaked in water

Table 2 Aluminium in meat, fruit and other foods (all concentrations in mg/kg fresh weight)

Food	Koivistoinen (3) 1980	Teraoka (4) 1981	MAFF (5) 1985	Greger (6) 1985	Delves (8) 1988
	Reference source				
Beef) Pork)	4	3	2	0.2	-
Eggs	-	-	-	-	<0.2
Butter	-	-	-	-	0.05
Margarine	-	-	-	-	0.2
Cheese	2-3	-	-	16	0.2
Apple	1	0.1	-	)	0.1*
Pear	<1	0.1	-	)0.4	-
Tea infusion	3	-	4	5	1-2

*in flesh, up to 4 mg/kg in skin

A recent paper published in Nature[9] reports a mean value of about 4 mg/kg for aluminium in tea. On the basis of the data in these tables, the dietary intake of aluminium would be expected to lie between 1 and 10 mg/day (assuming consumption of 1.5 kg of food per day).

Aluminium in milk and infant formula Information on cows' milk, infant formula and breast milk is presented in Table 3. The data on concentrations are presented in the order in which they were published with the oldest data at the top of the table. There appears to be a trend in the data, the more recent values tending to be lower than the older values. Once again any trend is likely to be due to improvements in analytical methodology rather than true differences between the actual concentrations of aluminium in milk and formula. For each of the three milks the most recent references - 8, 12, 14 - indicate similar aluminium concentrations; all are less than 0.1 mg/kg. In Japan[4] it was found that soya beans contained about 12 mg/kg of aluminium which is at the upper end of the concentration range reported for aluminium in vegetables. The indications are[13,14] that infant formula containing soya has elevated concentrations of aluminium: up to nearly 1 mg/kg on fresh weight basis.

Table 3 Aluminium in cow's milk, infant formula and breast milk (all on an "as consumed" basis)

Aluminium concentration mg/kg (reference in parenthesis)

Cow's milk	Infant formula	Breast milk
-	-	0.2-1 (10)
0.7 (3)	0.5 (3)	0.5 (3)
0.9 (4)	-	-
-	-	0.2-2.4 (11)
0.7 (6)	-	-
-	0.1-0.4 (1)	0.004 (1)
0.09 (12)	0.1-0.5 (12)	0.03 (12)
-	0.09-0.5 (13)	-
0.02 (8)	-	-
	0.02-0.1 (14)	-

(1) Freudlich, 1985
(3) Koivistoinen, 1980
(4) Teraoka, 1981
(6) Greger, 1985
(8) Delves, 1988
(10) Gueguen, 1971
(11) Iyengar, 1982
(12) Weintraub, 1986
(13) McGraw, 1986
(14) Baxter, 1988

2 NORMAL DIETARY INTAKE OF ALUMINIUM

Table 4 presents information on the dietary intake of aluminium in various countries. With the exception of the intake of 27 mg/day reported by Greger, all the intakes are less than 10 mg/day. Greger's estimate assumed that all cheese in the diet was processed and contained aluminium additives. In this instance about 75% of the dietary intake was from grains where baking powder was identified as a major source of aluminium. In the UK total diet the intake from cereals (equivalent to grains) was less than 2 mg/day, presumably because aluminium salts are less widely used in raising agents in the UK. The similarity between these estimates of intake is remarkable; such concordance is not normally seen for dietary intakes of other metals. The agreement is all the more surprising in view of the possible analytical problems referred to earlier. It seems reasonable to say that the average dietary intake of aluminium would be <u>about</u> 6 mg/day.

Table 4 Dietary intake of aluminium

Intake mg/day	Country	Reference
8	UK	MAFF, 1985 (5)
6	UK	MAFF, 1985 (5)
3	USA	Gorsky, 1979 (15)
5	USA	Greger and Baier, 1973 (16)
27	USA	Greger, 1985 (6)
6	USA	Gormian, 1970 (17)
7	Finland	Koivistoinen, 1980 (3)
4	Switzerland	Knutti,1985 (18)
5	Japan	Teraoka, 1981 (4)

Factors affecting intake

Aluminium from cookware One of the most common sources of additional dietary aluminium referred to in the literature is aluminium cookware. There can be no doubt that cooking acidic foods in aluminium vessels will increase the concentration of aluminium in the food. Cooking fruit, especially rhubarb, is a well-known way of cleaning up a saucepan. This cleaning action is in part achieved by the dissolution of aluminium metal. Various studies[19,6] concluded that aluminium from cookware makes only a modest contribution to dietary intake even under the worst case conditions. Recently interest in aluminium from cookware increased dramatically following publication of an article[20] which indicated that the presence of fluoride could accelerate the dissolution of aluminium into aqueous solutions of dilute organic acids and into foods. Subsequent research[21] indicated that for normal concentrations of fluoride in water - 1 mg/kg - fluoride did not accelerate the dissolution of aluminium into food. It now appears[22] that some of the original estimates of aluminium leaching in the presence of fluoride were in error. This must surely emphasise the need for ensuring that good quality control is used on all analyses but especially where the data are likely to arouse controversy and public anxiety: this caveat applies both to scientists producing and to journals publishing it.

The question of aluminium from cookware will re-emerge as a topic of interest at some time in the future. It is therefore instructive to examine the problem from a practical viewpoint. Imagine a family of 5 each of whom consumes 200 g/day of fruit boiled in an aluminium saucepan. A very high consumption rate has been chosen to ensure that sufficient aluminium is consumed in the food to have a major impact on the dietary intake of aluminium. On the basis of the admittedly erroneous published data[21], the aluminium concentration in the fruit after cooking could be 150 mg/kg. Since 1 kg of fruit has been cooked then the aluminium saucepan would lose 150 mg in weight. The weight of aluminium in contact with the food would be about 400 g. Consequently the maximum life of the saucepan would be about 7 years, that is the length of time before the saucepan had completely dissolved. As far as I am aware aluminium saucepans last a good deal longer than this. In this extreme example the extra intake of aluminium would be 30 mg/person/day, which is certainly elevated above normal but not dramatically so. If the consumption of stewed fruit was less extreme, say 1 kg/week or the aluminium concentration lower, say 20 mg/kg, then the additional aluminium intake would be about 4 mg/day which would mean that intakes were not much elevated above the average. Thus whilst cooking fruit in aluminium cookware could by some accounts cause a large increase in aluminium intakes it seems unlikely that it would happen in practice. In summary, you cannot have your aluminium saucepan and eat it.

Additives Aluminium-containing additives are used to perform various functions in food, for example sodium aluminium phosphate is used as a source of acid in raising agents and aluminium sodium silicate is used as an anti-caking agent, that is to prevent particles of food adhering to each other. In the USA it has been found[6] that some foods contain more than 1000 mg/kg of aluminium through the use of additives and that these additives make a major contribution, 20 to 25 mg/day, to the dietary intake of aluminium. However, other information (Table 4) indicates that these figures are overestimates, since the intakes of aluminium from whole diets were all less than 10 mg/day. MAFF is currently making an extensive survey of food additive usage in the UK and from this will derive information about intakes of additives including aluminium additives.

Antacids Antacids are a major source of aluminium intake for some individuals[23]. It is estimated that the maximum intake of aluminium from antacids lies in the region of 800 to 5000 mg/day. These intakes are 2 or 3 orders of magnitude greater than normal dietary intakes. Dietary intakes are most unlikely to reach such high values. This means that, for individuals consuming antacids, food is only a very minor source of aluminium intake.

Soil contamination The presence of soil in vegetables may increase aluminium concentrations to about 50 mg/kg as has already been demonstrated: this is at least one order of magnitude above normal values. The normal daily diet contains about 0.3 kg of vegetables; therefore if all these were contaminated with 0.1% soil on a fresh weight basis (about 1% on a dry matter basis) then this could contribute an extra 15 mg/day to the dietary aluminium intake. The impact of soil contamination on a grazing animal's intake of aluminium is likely to be very considerable because virtually the whole of the animal's diet is liable to soil contamination. At first sight this could have implications for the aluminium concentration in foods of animal origin. The diet of a cow inevitably contains some soil, regardless of whether the cow is eating silage or fresh grass. Cows' diets usually contain more than 1% w/w soil, say 5% is a typical value[24]. Assuming a cow consumes 10 kg of dry matter a day containing 0.5 kg of soil having an aluminium concentration of 50,000 mg/kg, then the intake of aluminium by a cow would be 25,000 mg/day (that is 25 g/day). Thus for a cow the intake of aluminium on a body weight basis would be about 50 mg/kg bw/day, whereas for a human it would be about 0.1 mg/kg bw/day. This 500 fold difference in intake of aluminium between cows and humans might lead to the conclusion that elevated levels of aluminium would be expected in milk or meat. However, this observation does not take account of the bioavailability of ingested aluminium. For completeness the intake of aluminium by a 3-month old infant receiving formula containing 0.05 mg/kg of aluminium would be about 0.01 mg/kg bw/day; this is an order of magnitude less than for an adult.

Bioavailability There is no doubt that hazards from metals in food are not solely related to exposure, that is what is taken in, but also what is taken up through the gut. The importance of this has already been demonstrated in the case of lead[25]. The major factor influencing aluminium uptake through the gut will be its solubility which in turn must be affected by both pH and the presence of complexing agents. Thus in man the pH of gastric juice is about 2, whereas in the rumen of a cow consuming forage it is between 6.2 and 6.7[26]. Thus all other things being equal it would be reasonable to expect that aluminium would be less available for gut uptake in a cow than in a man. Perhaps it is this which accounts for the fact that, despite the large ratio in intake of aluminium between cows and man, cows' milk contains about the same aluminium concentration as breast milk. By the same token people who consume aluminium-containing antacid do so for the specific purpose of increasing the pH of the gastric juices. Thus, although the intake of aluminium from antacids is more than 2 orders of magnitude greater than that from food, it would be wrong to think that this would be reflected in uptake of aluminium from the gut.

The foregoing comments are merely intended to highlight the complexity of assessing the effect of intake of aluminium upon body burden and it is this which is of greatest importance. There is a need for information on the absorption of aluminium by man and the factors which affect it.

CONCLUSIONS

Aluminium concentrations in commonly consumed foods are generally less than 10 mg/kg and average intakes of aluminium in 5 countries are about 5 mg/day. Aluminium concentrations in cows' milk, breast milk and infant formula are similar and are generally less than 0.1 mg/kg. The presence in food of aluminium from cookware is most unlikely to have a major effect on dietary intakes of aluminium. There is a need for information about aluminium absorption by man and the factors which affect it.

REFERENCES

1. M.Freundlich, C.Abitbol, G.Zilleruello and J.Strauss, The Lancet, 1985, September,527-529

2. J.A.Findlow, J.R.Duffield, D.A.Evans and D.R.Williams, Recueil des Travaux Chemiques des Pays-Bas, 1987, 106, (6-7), 403

3. P.Koivistoinen, Acta Agriculturae Scandinavica, 1980, Supplementum 22, 115-139

4. H.Teraoka, F. Morli and J.Kobayashi, J.Japanese Soc. of Food and Nutrition, 1981, 34 (3), 222-229

5. Ministry of Agriculture, Fisheries and Food, 1985 Food Surveillance Paper No 15, London, HMSO

6. J.L.Greger, Food Technology, 1985, May, 73-80

7. D.M.Sullivan, D.F.Kehoe and R.L.Smith, J Assoc.Off.Anal.Chem, 1987, 70(1), 118-120

8. T. Delves, 1988. ibid

9. S.J.Fairweather-Tait, G.R.Moore and S.E.Jemil Fatemi, Nature, 1987, 330, 213

10. L.Geguen, Ann.Nutr.Alim, 1971, 25, A335-A381

11. G.V.Iyengar, 1982, IAEA TEC Doc, Vienna

12. R.Weintraub, G.Hams, M.Meerkin and A.Rosenberg Arch.disease in childhood, 1986, 61, 914-916

13. M.McGraw, N.Bishop, R.Jameson, M.Robinson, M. O'Hare, C.Hewitt and J.P.Day, The Lancet, 1986, January, 1517

14. M.Baxter, A.Burrell and R.Massey. 1988. To be published in Food Additives and Contaminants

15. J.E.Gorsky, A.A.Dietz, H.Spencer and D.Osis, Clin. Chem. 1979, 25, 1739

16. J.L.Greger and M.J.Baier, Food Chem.Toxicol. 1979, 21, 473-477

17. A.Gormian. J.Am.Diet.Assoc. 1970, 56, 397-403

18. R.Knutti and B.Zimmerli. Mitt.Gibiete.Lebensm.Hyg. 1985, 76, 208-232

19. P.A.Mattson. Var. Foda, 1971, 33, 231

20. K.Tennkone and S. Wickramanayake. Nature, 1987, 325, (15 January), 202

21. J.Savory, J.R.Nicholson and M.R.Wills, Nature 237, 107-108

22. K. Tennakone and S. Wickramanayake. Environmental Pollution, 1982, 49, 133-143

23. A.Lione, Food Chem.Toxicol, 1983, 21, 103-109

24. I.Thornton and P.Abrahams. Sci.Tot. Environ, 1983 2, 287-2948

25. J.C.Sherlock, Env Geochem and Health, 1987, 9 (2) 43-47

26. D.Beaver, 1988 (Personal communication). Institute for Grassland and Animal Production

Aluminium in Infant Formulae and Tea and Leaching during Cooking

M.J. Baxter, J.A. Burrell, H.M. Crews, and R.C. Massey

MINISTRY OF AGRICULTURE, FISHERIES AND FOOD, FOOD SCIENCE DIVISION, QUEEN STREET, NORWICH NR2 4SX, UK

1] INTRODUCTION

Aluminium is the third most abundant element in the earth's crust and whilst it does not appear to serve any useful biochemical function its presence in the diet has not in general been considered detrimental to health. Very little of the ingested aluminium present in food is absorbed from the gastrointestinal tract and the large proportion is excreted in faeces[1]. In fact this non-permeability of the alimentary tract towards aluminium would appear to be essential to human well being. If the protection afforded by the intestine is circumvented a number of clinical problems may arise. This can occur during renal dialysis and if elevated levels of aluminium are present in the dialysis fluid then dialysis dementia and dialysis osteodystrophy may result[2]. This situation will be further exacerbated by the use of oral aluminium hydroxide as a phosphate binder in patients on renal dialysis[2]. Aluminium has also been suggested to be a causative factor in the aetiology of Alzheimer's disease[3,4] due to the occurrence of the metal in senile plaques and neurofibrillary tangles in brain tissue of patients with this condition. It should however be noted that the ultrastructure of the neurological disorders associated with Alzheimer's disease are somewhat different from those known to be induced by aluminium[5]. Nevertheless the suggestion that aluminium might be implicated in this major cause of senile dementia has focused attention on dietary sources of the metal. In particular there has been renewed interest in the literature on the aluminium content of foods cooked in aluminium saucepans[6] and its presence in tea infusions[7] and infant formulae[8]. These dietary sources of aluminium are discussed together with the need for more information concerning the relative gastrointestinal absorption of the metal from different foodstuffs.

2] LEACHING DURING COOKING

It is well established that cooking of acidic foods in aluminium saucepans causes leaching of the metal[9,10]. Greger et al[11] have reported that the aluminium content of tomatoes increased from 0.12mg/kg to 3.1mg/kg on cooking in an aluminium saucepan. Similarly Baxter et al[12] have shown that cooking a tomato homogenate (pH 4.4) resulted in an aluminium content of 3.3mg/kg as compared with 0.5mg/kg in the uncooked sample. When an even more acidic commodity such as rhubarb (pH 3.2) was cooked the concentration rose from 1.0mg/kg to 11.3mg/kg[12]. The greater leaching observed with more acidic foods arises from dissolution of the protective oxide layer on the surface of the aluminium saucepan. The large majority of foods are not acidic and consequently very little leaching of aluminium occurs. Thus Greger et al[11] have reported that the aluminium content of potatoes remained below 0.5mg/kg after cooking. In addition to pH considerations the degree of aluminium leaching may also be influenced by the age, surface topography and alloy content of the saucepan and variations of up to an order of magnitude have been observed between different saucepans[11,12,13]. The aluminium concentration will also increase during cooking as a result of evaporation. This is illustrated in Figure 1 in which a solution of citric acid (500ml, 1%) was boiled for 30 minutes in an aluminium saucepan without a lid being fitted[14]. During the course of this experiment the volume of citric acid reduced from 500ml to 48.5ml. The aluminium concentration increased from 2.4mg/kg to 304mg/kg whilst the total amount of aluminium in solution rose from 1.2mg to 14.7mg. These results serve to illustrate that the concentration of aluminium is more sensitive to substantial evaporation of water vapour during boiling than the absolute amount of the metal in solution. Interest in the leaching of aluminium saucepans has recently been heightened by Tennakone and Wickramanayake[15] who reported that the presence of 1mg/kg of fluoride increased aluminium dissolution by up to 3 orders of magnitude at pH 3.0. These findings have not been confirmed by subsequent workers[6,7,12] and indeed Tennakone and Wickramanayake have now[16] withdrawn their original paper. This serves to highlight the need for adequate analytical quality assurance in trace analysis, particularly in the case of ubiquitous analytes such as aluminium where the risk of contamination during analysis is high.

FIGURE 1: Effect of evaporation on aluminium concentration during boiling citric acid solution (500ml, 1%) in an aluminium saucepan.

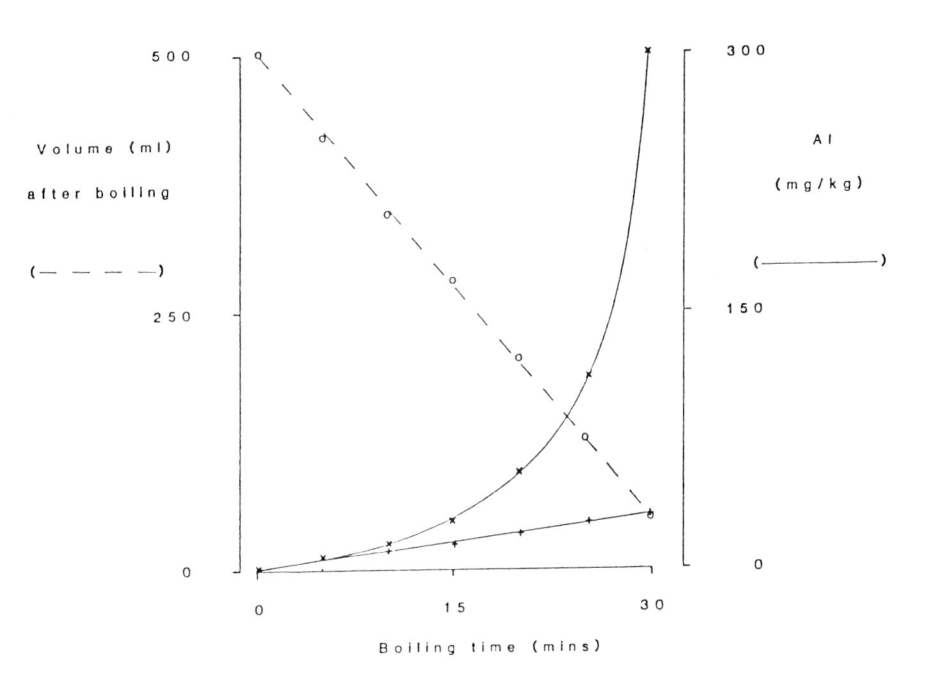

×, Al concentration in boiled solution

+, Al concentration in boiled solution after correction for weight loss caused by evaporation

3] ALUMINIUM IN TEA

Tea leaves are known to contain naturally high levels of aluminium with typical dry weight values of around 1000 µg/g[7,17]. In terms of dietary intake the concentration in tea infusions is of greater relevance as it is known that not all of the aluminium is extractable into hot water[7]. Literature values for aluminium in tea infusions are given in Table 1.

Table 1: Aluminium in Tea Infusions

Reference	Aluminium Content
Varo et al (1980)[18]	2 to 4mg/kg
Jackson (1983)[17]	1.9 to 3.9mg/kg
Coriat and Gillard (1986)[19]	40 to 100mg/kg
Fairweather-Tait et al (1987)[7]	2.7 to 4.9mg/l
Koch et al (1988)[20]	4.5 to 6.0mg/l
Baxter et al (1989)[21]	2.2 to 4.5mg/l

It is evident that with the exception of the data of Coriat and Gillard[19] there is a consensus that the aluminium concentration in tea infusions is in the range 2 to 6mg/l. Coriat and Gillard have not followed up their original letter to Nature and details of their analytical method and quality assurance procedures are not available. It therefore seems reasonable to assume that their results may have been in error. This supposition is further supported by the findings of Fairweather-Tait et al[7] who observed that some 26 to 39% of the aluminium in tea leaves was extracted during preparation of tea infusions which contained 2.7 to 4.9mg/l. If all of the aluminium had been extracted the maximum possible concentration in the infusion would have been 13.5mg/l. This is still a factor of 3 below the lowest concentration reported by Coriat and Gillard.

4] ALUMINIUM IN INFANT FORMULAE

The presence of aluminium in infant formulae has attracted interest in recent years as the immature gastrointestinal tract may be more permeable to aluminium than in the adult. This situation may be particularly exacerbated in infants suffering from uraemia[8]. Recent literature values for aluminium in infant formulae (wet weight) are shown in Table 2.

TABLE 2: ALUMINIUM IN INFANT FORMULAE*

Reference	Aluminium Content
a) Cows milk based formulae	
Jones (1985)[22]	<0.05 to 0.11mg/kg
Hewitt et al (1987)[23]	0.04 to 0.37mg/l
Koo et al (1988)[24]	0.02 to 0.57mg/l
Baxter et al (1989)[21]	0.03 to 0.20mg/l
b) Soya based formulae	
Jones (1985)[22]	0.41 to 2.80mg/kg
Hewitt et al (1987)[23]	0.42 to 1.33mg/l
Koo et al (1988)[24]	0.46 to 2.35mg/l
Baxter et al (1989)[21]	0.64 to 1.34mg/l

*, all results are expressed in terms of concentration in infant formula made up ready for consumption according to manufacturers instructions.

Although concentrations are low in all samples it is evident that soya based infant formulae contain higher amounts of aluminium than the cows milk based samples. The reasons for this have yet to be established. A number of possibilities exist including i) contamination during manufacture, ii) aluminium present in vitamins and nutrient elements which may be added in higher amounts to the soya formulae and iii) higher levels of aluminium in soyabeans as compared with cows milk.

The third seems likely to predominate, ie, in view of the non-permeability of the mammalian adult gastrointestinal tract towards aluminium it seems probable that levels of the metal may well be substantially lower in cows milk as compared with materials of plant origin. In this context it is interesting to note that the concentrations of aluminium in cows milk (0.024 to 0.029mg/l, Koo et al 1988[24]; 0.09mg/l, Weintraub et al 1986[25]; 0.02 to 0.06mg/l, Pennington and Jones 1988[26]) are significantly lower than the 13 to 14 mg/kg in soy meal reported by Varo et al 1980[27]. The data for cows milk infant formulae reported by Jones 1985[22] (<0.05 to 0.11mg/kg) and Baxter et al 1989[21] (0.03 to 0.20mg/l) are very similar to the above results for cows milk itself. This implies that for these particular samples at least aluminium was not present in significant amounts in the other nutrients added to the infant formulae. It also demonstrates that contamination by aluminium during manufacture and storage was not a problem. Somewhat higher levels of up to 0.57mg/l and 0.37mg/l respectively were detected in some of cows milk infant formulae analysed by Koo et al 1988[24] and Hewitt et al 1987[23]. It is therefore apparent that cows milk based infant formulae may on occasion contain somewhat higher levels of aluminium than are naturally present in cows milk and also for that matter in human breast milk (<0.005 to 0.045mg/l, Koo et al 1988[24]; 0.03mg/l, Weintraub et al 1986[25]). In general however the levels are similar to those in cows milk and lower than the amounts in soya based products.

5] CONCLUSIONS

Much of the recent interest over aluminium in food, beverages and water stems from the suggestion[3,4] that it may be a factor in the aetiology of Alzheimer's disease. In order that this putative link may be objectively investigated it is essential that accurate data are available on human exposure to aluminium and the levels at which the metal is present in the diet. This is not a straightforward task in view of the potential for contamination by extraneous aluminium during analysis and the technical difficulties of measuring low concentrations of the metal. However it is apparent from the foregoing sections that reliable values now exist particularly for important commodities such as tea infusions and infant formulae. Unfortunately there are very few certified reference materials with which individual laboratories can check the accuracy of their results. Those that do exist include riverine water (23.5ug/l, SLRS-1, National Research Council Canada, Ottawa, Canada), citrus leaves (92mg/kg, SRM 1572, National Bureau of Standards, Washington DC, USA) and tea leaves (775mg/kg, No.7, National Institute for Environmental Studies, Tsukuba Ibaraki, Japan). These materials are at concentrations and in matrices which are atypical of the majority of foods.

Total diet studies reveal that the average daily intake of aluminium from foodstuffs in the UK is 6mg[28]. Considerably greater amounts of up to several grammes per day may be consumed in the form of aluminium containing antacid preparations[29]. The absolute concentration of aluminium consumed in foods and medications is in fact only one of the factors that influences human exposure. Of equal or possibly greater relevance is the proportion and subsequent fate of dietary aluminium that is absorbed from the gastrointestinal tract. It is established that only a small percentage of ingested aluminium is absorbed[1]. However accurate information concerning the actual amount absorbed is currently unavailable. Similarly there are no reported investigations into whether aluminium is more readily absorbed from some dietary commodities as compared with others. In the case of lead[30] it is known that some 40 to 50% of an aqueous challenge is absorbed by fasting subjects, 14% when taken with tea and only 7% when the lead is ingested as part of a meal. The proportion of dietary aluminium absorbed is much lower than this. However because of the lower percentages involved it is conceivable that the absolute amount absorbed may vary by orders of magnitude depending on the composition of the beverage or food. There appears to be a dearth of information concerning the chemical forms of aluminium in the gastrointestinal tract. That speciation is important is underlined by the observation that the presence of citrate significantly enhances absorption[31] and this is presumably a consequence of the formation of neutral, membrane-permeable aluminium citrate complexes[32,33]. One of the major obstacles in studying the factors affecting the absorption of aluminium is the absence of a suitable radio-isotope. A partial solution to this problem has been devised by Farrar et al 1988[34] who have employed ^{67}Ga as a model for aluminium. Intestinal absorption of gallium in the rat was enhanced by maltol (a flavouring additive) in fasted animals but no effect was observed in fed controls. Interestingly both gallium and aluminium are known to form highly stable, neutral and lipophilic complexes with maltol[35].

The use of inductively coupled plasma - mass spectrometry (ICP-MS) as a highly sensitive and selective HPLC detector offers the potential of examining metal speciation at µg/kg levels[36,37]. We are employing this technique to examine the chemical form of aluminium in foodstuffs and Figure 2 shows a size exclusion chromatogram[38] of a tea infusion with the ICP-MS monitoring 27 amu. The chromatogram reveals the presence of two peaks whose retention times (22.3 and 23.6 minutes) correspond to molecular weights of 13000 and 6700 daltons respectively. The identities of these aluminium binding species have yet to be established but it seems likely that they may be polyphenolic thearubigens[39]. An assessment of the factors that

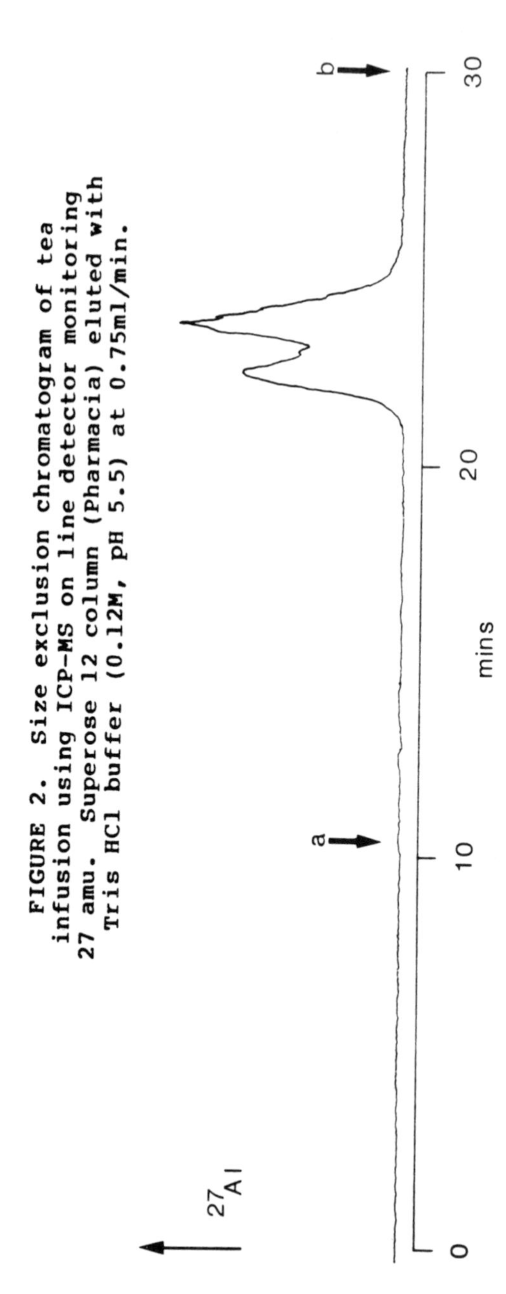

FIGURE 2. Size exclusion chromatogram of tea infusion using ICP-MS on line detector monitoring 27 amu. Superose 12 column (Pharmacia) eluted with Tris HCl buffer (0.12M, pH 5.5) at 0.75ml/min.

a,b: expected retention times of totally excluded species of >300000 daltons (a) and totally permeating, non-adsorbed species of <1000 daltons (b).

affect the absorption of these species is of considerable interest as tea is likely to be the major dietary source of aluminium for regular consumers and Koch et al 1988[20] have recently demonstrated that urinary levels of the metal increase after drinking tea. We are currently examining the effects of simulated gastric and intestinal digestion on the chemical forms of aluminium in tea and the findings will be reported elsewhere.

REFERENCES

1. J.L.Greger and M.J.Baier, Food and Chemical Toxicology, 1983, 21, 473-477.

2. W.K.Stewart, ibid, 1989.

3. J.A.Edwardson, ibid, 1989.

4. C.N.Martyn, C.Osmond, J.A.Edwardson, D.J.P.Barker, E.C.Harris and R.F.Lacey, The Lancet, 1989, 59-62.

5. O.Bugiani and B.Ghetti, Neurological Ageing, 1982, 3, 209-222.

6. J.Savory, J.R.Nicholson, and M.R.Wills, Nature (London), 1987, 327, 107-108.

7. S.J.Fairweather-Tait, R.M.Faulks, S.J.A. Fatemi and G.R.Moore, Human Nutrition: Food Sciences and Nutrition, 1987, 41F, 183-192.

8. M.Freundlich, C.Abitol, G.Zilleruelo and J.Strauss, Lancet, 1985, September 7, 527-529.

9. C.F.Poe and J.M.Leberman, Food Technology, 1949, 3, 71-74.

10. G.A.Trapp and J.B.Cannon, New England Journal of Medicine, 1981, 304, 172.

11. J.L.Greger, W.Guetz and D.Sullivan, Journal of Food Protection, 1985, 48, 772-777.

12. M.J.Baxter, J.A.Burrell and R.C.Massey, Food Additives and Contaminants, 1988, 5, 651-656.

13. O.Knoll, H.Lahl, J.Bockmann, H.Hennig and B.Unterhalt, Trace Elements in Medicine, 1981, 304, 172-173.

14. M.J.Baxter, J.A.Burrell and R.C.Massey, unpublished results.

15. K.Tennakone and S.Wickramanayake, Nature (London), 1987, 325, 202.

16. K.Tennakone and S.Wickramanayake, Nature (London), 1987, 329, 358.

17. M.L.Jackson, The Science of the Total Environment, 1983, 28, 269-276.

18. P.Varo, M.Nuurtamo, E.Saari and P.Koivistoinen, Acta Agriculturae Scandinavica, 1980, Supplementum 22, 127-139.

19. A.-M.Coriat and R.D.Gillard, Nature (London), 1986, 321, 570.

20. K.R.Koch, M.A.B.Pougnet, S.DeVilliers and F.Monteagudo, Nature (London), 1988, 333, 122.

21. M.J.Baxter, J.A.Burrell and R.C.Massey, Food Additives and Contaminants, 1989, in press.

22. J.W.Jones, unpublished data quoted by J.A.T.Pennington, Food Additives and Contaminants, 1988, 5, 161-232.

23. C.D.Hewitt, M.O'Hara, J.P.Day and N.Bishop. In: Trace Element-Analytical Chemistry in Medicine and Biology, edited by P.Bratter and P.Schramel, Walter de Gruyter and Co. (Berlin), 1987, 4, 481-488.

24. W.W.K.Koo, L.A.Kaplan and S.K.Krug-Wispe, Journal of Parenteral and Enteral Nutrition, 1988, 12, 170-173.

25. R.Weintraub, G.Hams, M.Meerkin and A.R.Rosenberg, Archives of Disease in Childhood, 1986, 61, 914-916.

26. J.A.T.Pennington and J.W.Jones. Aluminium in Health, a Critical Review, edited by H.J. Gitelman, Marcel Dekker (New York), 1988, in press.

27. P.Varo, M.Nuurtamo, E.Saari and P.Koivistoinen, Acta Agriculturae Scandinavica, 1980, Supplementum 22, 37-55.

28. Ministry of Agriculture, Fisheries and Food, 1985, Food Surveillance Paper No 15, London, HMSO.

29. A.Lione, Pharmaceutical Therapeutics, 1985, 29, 255-285.

30. M.J.Heard, A.C.Chamberlain and J.C.Sherlock, Science of the Total Environment, 1983, 30, 245-253.

31. P.Slanina, W.Frech, L.-G.Ekstrom, L.Loof, S.Slorach and A.Cedergren, Clinical Chemistry, 1986, 32, 539-541.

32. J.A.Findlow, J.R.Duffield, D.A.Evans and D.R.Williams, Recueil des Travaux Chimiques des Pays-Bas, 1987, 106, 403.

33. R.B.Martin, Clinical Chemistry, 1986, 32, 1797-1806.

34. G.Farrar, A.P.Morton and J.A.Blair, Food and Chemical Toxicology, 1988, 26, 523-525.

35. M.M.Finnegan, T.G.Lutz, W.O.Nelson, A.Smith and C.Orvig, Inorganic Chemistry, 1987, 26, 2171-2176.

36. J.R.Dean, S.Munro, L.Ebdon, H.M.Crews and R.C.Massey, Journal of Analytical Atomic Spectrometry, 1987, 2, 607-610.

37. H.M.Crews, J.R.Dean, L.Ebdon and R.C.Massey, Analyst, 1989, in print.

38. H.M.Crews and R.C.Massey, unpublished results.

39. Kirk-Othmer, Encyclopedia of Chemical Technology (John Wiley and Sons), 1983, 22, 628-644.

inhibited by the presence of sugars and fatty substances that have an extinguishing effect on the corrosive processes.

It must be mentioned here that there are at least two methods commonly used to increase the thickness of the natural oxide layer on aluminium. One is the so-called Bohmitizing process. By putting aluminium into de-ionised boiling water or steam a surface layer of Al(OOH) Bohmit of a thickness up to 2 µm is formed. Another method to get up to 20 µm thick oxide layers on aluminium is the widely used anodic oxidation in diluted sulphuric or oxalic acid. This anodic oxide layer can be produced in a variety of selected grades of hardness and wear resistance with good corrosion resistance within the pH range of 4 and 8.5. Kitchen utensils and pots of aluminium sometimes have such decorative anodic oxidised surfaces. Over the last ten years an increasing portion of aluminium utensils used in the kitchen as pots and pans is produced with a coated surface. The coating is either a sintered fluoro-carbon layer (Teflon) or stainless steel. Three layer sandwich type metal structures of stainless steel + aluminium + stainless steel have gained a large market share for pots and pans in many countries. The reason for using aluminium in steel clad hollow ware is its superior heat conductivity and low specific weight.

Aluminium used in packaging, plain or coated

In packaging aluminium is used either plain or converted. There are three major applications of plain aluminium:
- household foil
- foil containers
- wrappings (either paper laminated or plastic coated for sealing) eg for chocolate, effervescent tablets.

Foil containers are used for chilled and frozen food, bakery products, baked meat pies and for prepared takeaway food. Greger et al[1] have investigated the aluminium content of foods before and after being frozen, refrigerated and cooked or baked in aluminium foil or trays (Table 1). Their results show a very small increase in the content of aluminium in the food, which can be neglected comparing with the differences which can be found in the natural aluminium content of food.

Table 1 Aluminium content of foods before and after being frozen, refrigerated and cooked in aluminium foil/trays

Food	Uncooked food µg Al/g wet wt	Prepared food µg Al/g wet wt
Beef	0.30	0.33
Flounder	0.55	0.70
Turkey	0.32	0.38
TV dinner (mashed potatoes)	0.18	0.97

Besides these three applications of plain aluminium the vast majority of aluminium in packaging is used in converted form. Converted means that the packed goods do not get in contact with the aluminium itself but rather with an intermediate layer of lacquer, plastic, paper, or cardboard covering the aluminium (Table 2). These coatings serve as protective layers, for example as can coatings or as heat sealing means to produce tightly sealed lids on blisters, cups, containers or composite cans, to close pouches, composite tubes or wrappings.

Special consideration has been given by the health authorities to the constituents of organic coatings. Most countries of the EEC and also the USA have so called 'positive lists'. These positive lists contain all monomers and additives allowed to make resins to be used in contact with foodstuff.

In the UK and Germany these positive lists are recommendations. The producer of packaging material must guarantee to the user that these recommendations are followed, and the recommendations are constantly kept up to date with the science. In addition to the 'positive lists' in some countries there are limits of global migration and limits of specific migration or organic materials in contact with food.

Table 2 Converted aluminium for packaging

Conversion	Resin	Packaging examples
Protective stove-lacquering	Epoxy-phenolic, Vinyls	Containers Cans
Heat seal lacquers	Vinyls	Push through Lids
Extrusion coatings	Polyolefins	Brik pack Laminate tube
Laminated plastics, paper or cardboard	Polyurethanes	Pouches Foilboard

2 REGULATIONS ON ALUMINIUM AND ALUMINIUM ALLOYS USED IN CONTACT WITH FOOD

Most countries of the world, including the European Community, do not have specific requirements for light metal alloys in contact with food. The reason for this is the virtually non-existent risk commonly associated with aluminium and its alloys from a toxicological point of view.

However, as there is an increased use of this metal in many areas of our daily life discussions and investigations on its toxicological aspects are now taking place.

Only France has passed regulations, by publishing a law in 1987 for products made of aluminium and its alloys for direct contact with food and beverages for human and animal use. This does, of course, not concern packaging materials which are coated. The permitted alloying elements and their maximum amount are listed in Table 3, the chief constitutent must be aluminium. It is expected that regulations of this kind will follow in other countries and the EC.

Table 3 French Regulation (August 1987). Limits of alloying elements in Al alloys used in contact with food

Si	$\leq$ 13.5%
Mg	$\leq$ 11.0%
Mn	$\leq$ 4.0%
Ni	$\leq$ 3.0%
Fe	$\leq$ 2.0%
Cu	$\leq$ 0.6%
Sb	$\leq$ 0.4%
Cr	$\leq$ 0.35%
Ti	$\leq$ 0.3%
Zr	$\leq$ 0.3%
Zn	$\leq$ 0.25%
Sr	$\leq$ 0.2%
Sn	$\leq$ 0.1%
As, Ta, Be, Tl, Pb each	$\leq$ 0.05%
in total	$\leq$ 0.15%

Source: Ministere de l'economie, des finances et de la privatisation, Paris, 1987

3 THE USE OF ALUMINIUM AS A FLEXIBLE, SEMI-RIGID OR RIGID PACKAGING MATERIAL

Properties of aluminium as a packaging material

As mentioned earlier, aluminium has found an extensive use in packaging. This is not surprising considering its packaging related properties (Table 4).

Increase in use of packaging materials

A comparison of packaging materials used in Germany in 1980 and 1986 (Fig 2) shows no increase in tonnage for paper and glass but a significant increase in the use of plastics and metals. 'Metals' includes steel and aluminium. The use of aluminium as packaging material in the UK over the last 6 years is shown in Fig 3. The total annual consumption rose from about 95,000 t to 130,000 t with a more significant increase in the use of aluminium for cans compared to foil.

One of the most important properties of Al in packaging is its barrier function against the penetration of oxygen, moisture, light, odours and grease. As this barrier function is exercised even by very thin aluminium foil, there is an increased tendency towards the use of aluminium foil in packaging. By the combination of thin layers of various materials with foil a remarkably reduced material input can be achieved for a given volume of goods to be packed. This type of semi-rigid and flexible packaging fits well into the trend of all municipalities in industrialised countries to cut back volume and weight of waste from packaging.

The increase of Al-foil consumption for packaging in Europe over the last 15 years is shown in Fig 4. The growth comes mostly from foil in thickness below 7 µm. Aluminium foil consumption for packaging is in close relation to the GNP of a country. The higher the standard of living is, the more packaged products and packaging materials are used.

Table 4 Packaging related properties of aluminium

BARRIER against gases, moisture, flavours, grease, oil, light, non-degenerating

TASTELESS corrosion resistant

CONDUCTIVE thermally and electrically, no static charges

GOOD MECHANICAL PROPERTIES at low and elevated temperatures, no change as a function of time

GOOD FORMABILITY at ambient temperature, deep drawing, stamping, dead folding

EASY COMBINATION with other materials, eg paper, plastics

DECORATIVE APPEARANCE and compatible with printing processes

LOW SPECIFIC WEIGHT 2.7 g/cm^3

CAN BE RECYCLED

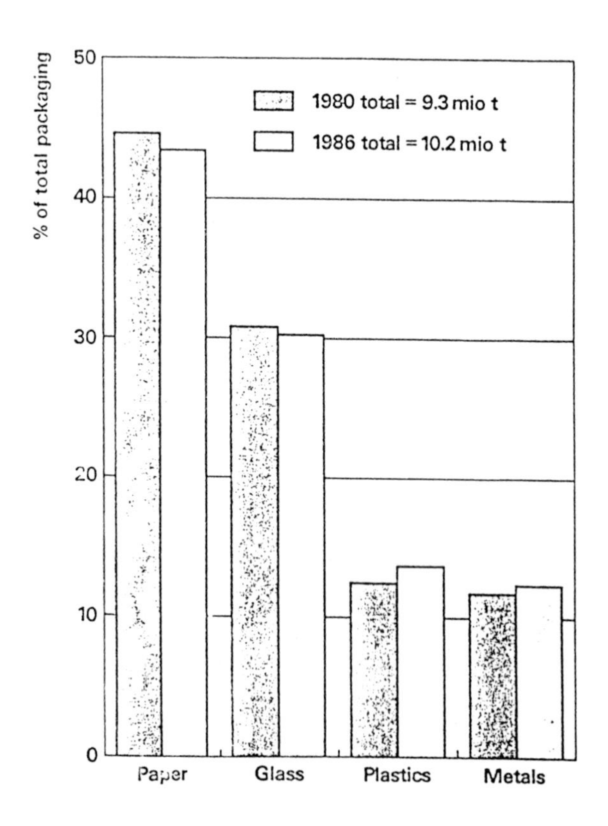

Figure 2 Packaging materials produced in Germany (Source: RG Verpackung im RKW, Düsseldorf)

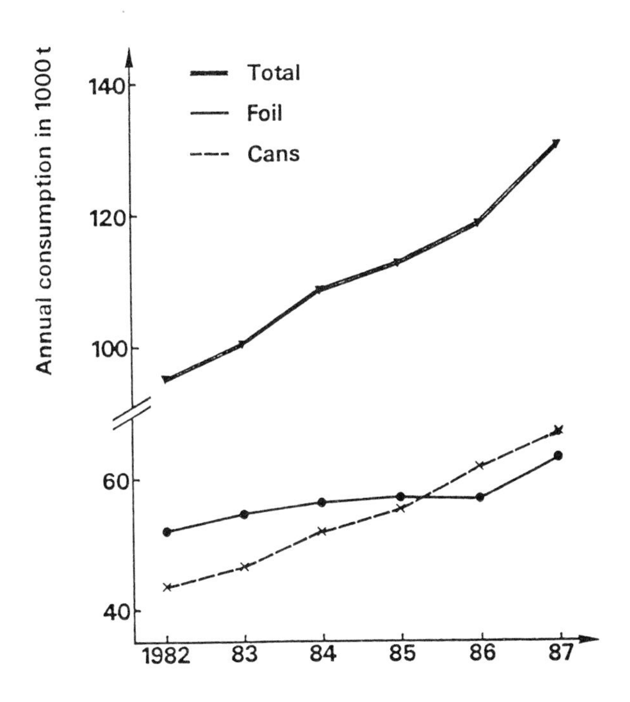

Fig 3 Aluminium consumption for packaging in UK (Source: Alusuisse UK)

4 POTENTIAL FOR MIGRATION OF ALUMINIUM INTO FOODSTUFFS

Foodstuff in contact with plain aluminium

Foodstuff with neutral activity and low salt content does not significantly attack aluminium with its natural oxide layer. Aluminium behaves indifferently in contact with milk, natural milk products, edible fats, and oil at ambient temperature. Glucose, protein and pectin act as inhibitors in products containing fruit acids. Investigations with the preparation of food in plain aluminium utensils at cooking and boiling temperatures cover a more stringent application than is usually found in packaging where the product is at ambient temperature and the surface of the aluminium is usually coated.

J. L. Greger and co-workers[1] in the US have analysed aluminium levels in foods cooked in plain aluminium pans (Table 5). They found that the amount of

The white wine was Riesling Sylvaner from the Kanton Geneva vintage 1983. All the characteristics of the wine were analytically examined. The pH value was 3.66 which is quite low. Cans and bottles were filled under defined conditions of CO_2 atmosphere and head space.

Cans and bottles were stored at 20°C ± 3°C. Taste and odour evaluation and analytical investigations of any changes were done after 1, 7, 26, 52 and 100 weeks. The changes of metal contents for Al, Fe, Sn, Na, K and Cu, analysed by atomic absorption spectrometry, are presented in Table 8. The relatively high concentrations of iron and tin in wine stored in tin cans come from little corrosion spots at the welded seam of the can body, which was covered with a 2 component lacquer. The aluminium content in the glass bottles and the tin cans was only slightly lower compared to the aluminium content in the aluminium can.

Recently another investigation was completed in our packing laboratory on the influence of different container materials and storage time on the content of aldehydes, ketones and metal contents in apple juice[4]. Aldehydes and ketones are very important odour and taste components in fruits and juices, which influence the quality of such products.

The results of the metal contents analysed in the original apple juice and after storage for 22 months in different containers are of interest (Table 9). This extremely long storage time of 22 months was chosen for the purpose of evaluating extreme conditions which, of course, do not occur in practical use. After 22 months the aluminium content has only slightly increased because of tiny spots of corrosion at the formed area of the lid.

Table 8 Content of various elements in white wine after two years storage in different containers (mg/l)

Element	Glass bottle 200 ml	Al can/ ALUFIX 160 ml	Al can/ ring pull 160 ml	Tin can 160 ml
Al	1.9	2.1	2.6	1.5
Fe	3.9	4.0	4.0	317.0
Sn	1.5	1.5	1.5	3.1
Na	12.0	13.0	12.0	14.0
K	1.0	1.0	1.0	1.0
Cu	0.2	-	0.1	-

Table 9 Content of Fe,Sn,Al in apple juice in different containers (mg/l)

Elements	Tin can 330 ml 2-piece lacquered		Tin can 160 ml 3-piece lacquered and seam coated		Al can 330 ml 2-piece lacquered		Al can 150 ml 2 piece lacquered	
	0	22	0	22	0	22	0	22
	months		months		months		months	
Fe	0.9	3.9	0.9	26	0.9	0.9	0.9	0.9
Sn	<1.0	4.0	<1.0	27	<1.0	<1.0	<1.0	<1.0
Al	1.3	1.3	1.3	1.3	1.3	9.4	1.3	5.2

ALUMINIUM MIGRATION DURING FOOD STORAGE AND PREPARATION A PERSPECTIVE VIEW

The previous section has shown that measurements of the amount of aluminium transferred from cookware and packaging materials indicate a maximum addition of 10 mg/kg. However, these data need to be placed in perspective against the other sources of this metal in the human diet.

Sherlock, in a previous chapter points out that the aluminium content of most foodstuffs lies in the range of 0.1 to 10 mg/kg. In addition aluminium is used extensively in a variety of food additives (Table 10), approximately 2000 tonnes/annum being used for this purpose in the USA[6].

The human dietary intake of aluminium from both these two sources and water has been estimated by a number of workers (Table 11). In addition, aluminium compounds continue to be used extensively in medical preparations, largely with antacid products and buffered analgesics. This use can lead to an intake of 5 g/day, which is orders of magnitude greater than any intake from either the diet or from migration during packaging or cooking.

Table 10 Aluminium in food additives

- Aluminium sulphates - pH control, firming, separation and filtration, colour control
- Bentonite - separation and filtration, clarifying, flocculating, colour control
- Sodium aluminium phosphate basic and acidic - leavening agent, stabilizer, emulsifier
- Sodium aluminium silicate - moisture control, appearance control

Table 11

- Daily intake of aluminium by normal diet	
- Food Surveillance Paper No 15, London	5-7 mg
- C. Schalatter, A. Kanzig, Zurich 1985	7-25 mg
- U. Candrian, Zurich, 1985	4-100 mg
- Vettorazzi 1987/J.L.Greger 1985	10- 50 mg
- M.E.De Broe, Antwerp 1987	20-100 mg

CONCLUSIONS

Aluminium is used in packaging in plain and converted form. Transfer from plain aluminium in packaging (chocolate wrap, household foil, foil containers) is in the range of 0-1 ppm for normal food which is of neutral reactivity with addition of sugar, pectin and fats which act as inhibitors against the attack of the aluminium.

From plain aluminium kitchen utensils and appliances aluminium contents up to 10 ppm can be transferred when foods of low pH are cooked. The same amount may migrate from stove lacquer coated aluminium cans into acidic beverages after extended storage. No aluminium transfer whatsoever is measurable from the mostly used converted aluminium packaging materials coated with thermoplastics and paper.

Compared with the variability of the natural aluminium content in food and beverages these amounts migrating into food products and beverages from aluminium packaging material are negligible. Usually the normal human body's largest intake of aluminium comes from vegetables and legumes, food additives, antacid products, analgesics and toiletries/cosmetics.

REFERENCES

1 J.L.Greger, W. Goetz, D. Sullivan. Journal of Food Protection Vol 48, Sept 1985

2 Ch. Schlatter, A. Kauzig. Untersuchungsbericht des Instituts fur Toxikologie der ETH und Universitat Zurich, Schwerzenbach 1985

3 P Durr, S Bloeck. Deutsche Lebensmittel-Rundschau H. 12 1987, S.385

4 S. Bloeck, A.Kreis and O. Stanek. Alimenta 2, 1988, S.23-28

5 U. Candrian. Mitt. Gebiete Lebensm.Hyg.,Band 76 (1985) S.570-608

6 J.L.Greger. Food Technology, May 1985, S.76

7 Prof. M.E. De Broe et al. University of Antwerp 1988

8 Skalsky H.L., Carchman R.A. J.Amer.Coll.Toxicol.2,1983

Subject Index

A4 amyloid,
  deposits: 26
  fibrils: 30
  peptide: 29
  protein, precursor: 23, 30, 32, 33
AAS; *see Atomic absorption spectrophotometry*
Absorption excluders: 1
Absorption,
  gastrointestinal barrier to: 26
  intestinal, of phosphate: 28, 53
  of dietary Al: 83
  of phosphate from the gut: 13
Accumulation, of Al: 21
Accuracy, of analyses of Al in foods: 60
Acetylcholine, loss of: 22
Additives, in food: 63, 73
Administration, oral,
  of Al: 12
  of aluminium hydroxide: 12, 13
Al^{3+}, complex with citrate: 42
$[Al(H_2O)_6]^{3+}$ cation, hydrolysis of: 41
Aldehydes,
  as odour components: 98
  as taste components: 98
Aluminate anion: 41
Aluminium,
  accumulation at the calcification front: 8
  accumulation in brain: 3, 7

Aluminium (contd.)
  accumulation in tea: 56
  accumulation of: 21
  administration of, oral: 12
  analysis in foods, accuracy: 60
  and Alzheimer's disease: 37, 38
  as ferrioxidase inhibitor: 44
  as packaging material: 93
  binding by phosphate: 44, 48
  bioavailability of: 65
  chemistry of: 41
  colorants: 56
  concentration, in plasma: 10
  concentration in serum: 10
  consumption for packaging in UK: 95
  contact with food: 88
  contamination, inadvertant: 16
  content, of dialysis fluid: 7
  content, of cerebral cortex: 24
  content, of hippocampus: 24
  deposition, on senile plaque core: 26, 32
  determination, in biological materials: 52
  determination, in foods: 52
  dietary, intake of: 71, 72
  dietary, absorption of: 83
  dissolution of, from cooking utensils: 5, 72
  dust, inhalation of: 21
  effect of fluoride on dissolution of: 72, 78
  excretion of, in urine: 65

Aluminium (contd.)
excretion of normal daily intake: 13
foil, consumption in Europe: 96
foil, in packaging: 93
from drinking water: 37
in blood plasma: 8
in bone biopsies: 53
in canned apple juice: 99
in cereals: 69
in cerebrospinal fluid: 53
in coffee: 97
in cows' milk: 70, 82
in dialysis fluid: 8
in dust entering the lungs: 4
in food additives: 99
in food, significance of: 63
in foods: 56, 68
in infant formulae: 70, 77, 81
in sea water: 17
in serum: 54
in serum, by ETA-AAS: 55
in soy meal: 82
in tea: 5, 77, 80
in vegetables: 69
in water sources: 38
in water supplies: 32, 97
ingested, bioavailability of: 74
intake from normal diet: 100
intake via inhalation: 48
intake, oral: 7
intra-cellular: 44
leaching, during cooking: 77, 78
levels, in foodstuffs: 4
metal: 88
migration, into foods: 95, 99
pathway of: 42
plain pans, Al levels of foods cooked in: 95
plasma concentrations of: 2
role in enzyme activity: 17
saucepans: 77
solubility of oxide on: 89
tangles in rabbits induced by: 24
toxicity, chronic: 53
toxicity to fish: 42

Aluminium (contd.)
toxicity to trees: 42
transfer from dialysis fluid to patient: 9
transport of: 48
uptake by coffee: 97
uptake by the brain: 26
uptake, via gastrointestinal tract: 5, 74, 75
urinary levels of: 85
world-wide consumption of: 2
Aluminium hydroxide,
administration, oral: 12, 13
at acid pH: 14
soluble, molecular form of: 15
therapy: 6, 52
Aluminium oxides: 1
Aluminium phosphates: 1, 56, 65
Aluminium silicates: 1
Aluminium sodium silicate: 73
Aluminium sulphate: 37, 99
Aluminon: 7
Aluminosilicate: 40, 41, 43, 44, 46, 47, 56, 65
amorphous: 26
deposits: 32, 33, 48
Alzheimer's disease: 20, 21, 22, 23, 24, 28, 29, 32, 33, 37, 38, 46, 68, 77
presenile: 24, 29
Amygdala: 28
Amyloid fibrils: 23, 29
β-Amyloid protein: 22, 23
Amyotrophic lateral sclerosis-parkinsonism-dementia complex: 21
Anaemia: 15, 40, 44
microcytic: 53
Antacid, action: 13
Antacids: 1, 3, 73
Antidiarrheals: 3
Antiphosphate absorber: 3
Apple juice: 98
canned, Al content: 99
canned, Fe content: 99
canned, Sn content: 99
Apple sauce: 96
Apples: 57, 60, 70
Atomic absorption, signals: 58

Atomic absorption spectrometry: 7, 52, 53, 98
flameless: 97
Atomisation, electrothermal: 53
Autoradiography: 27

Barrier, blood-brain: 27, 48, 68
Beans: 96
Beef: 70, 91, 96
Bentonite: 99
Bicarbonate: 47
Bioavailability, of Al: 65, 74
Biological materials, determination of Al in: 52
Biological properties, prediction of: 17
Blood-brain barrier: 27, 48, 68
Blood, monitoring of Al levels: 10
Blood plasma: 17
Al in: 8
Body burden: 10, 12, 40
Bohmitizing process: 90
Bone: 56
biopsies, Al in: 7, 53
disorders: 7, 53
Bottle feeds, for premature babies: 5
Brain,
Al accumulation in: 3, 7
Al uptake by: 26
psychomotor defects in: 16
Si uptake by: 26
Butter: 53, 58, 60, 63, 70

Cabbage: 69
Calcification front, Al accumulation at: 8
Calcium, in tangle-bearing neurones: 21
Carrots: 57, 61, 69
Celery: 69
Cereals, Al in: 69
Cerebral cortex: 27, 29
Cheese: 60, 62, 63, 70
Chicken: 96
Chromatography, size exclusion: 84
Chromium: 16
Citrate: 41, 44, 47, 65
complex with Al^{3+}: 42
Citric acid: 78
Coatings, organic: 91
Coffee: 57, 58, 64
Al in: 97
powder: 57, 63
uptake of Al by: 97
Complexes,
aluminium citrate: 83
membrane-permeable: 83
Consumption,
of Al foil in Europe: 96
of Al for packaging in UK: 95
Contamination, of protein infusion fluids: 52
Cookware, Al from: 72, 95
Cows' milk: 74
Custard, powder: 63, 64

DFO; *see Desferrioxamine*
Dairy products: 69
Dementia: 38, 53
dialysis: 3, 16, 40, 68, 77
senile: 21, 77
Density, of transferrin receptor: 47
Deposits, of aluminosilicate: 33, 48
Desferrioxamine: 12, 41
Dialysis,
dementia: 3, 16, 40, 68, 77
dementia syndrome: 6
encephalopathy: 30, 40
fluid, Al content: 7, 8
osteodystrophy: 77
renal: 2, 5, 77
Diet,
and Al in foods: 68
normal, Al intake from: 100
Disease,
Alzheimer's: 20, 21, 22, 23, 24, 28, 29, 32, 33, 37, 38, 68, 77
cause of: 48
consequence of: 48
Huntington's: 20
motor neurone: 20
Parkinson's: 20
renal: 6
Disorders,
neurodegenerative: 20
of bone: 53

Down's syndrome: 23, 26
Dust, Al in: 4

Eggs: 63, 64, 70, 97
Egg yolk: 45
Electron microprobe X-ray micro-analysis: 24
Elephantiasis: 48
Emission spectroscopy: 53
Encephalopathy: 6, 21
   dialysis: 30, 40
Entry of Al, via nasal-olfactory system: 48
Enzyme activity, role of Al in: 17
Epidemiological studies: 33
Epidemiology: 37
Epilepsy: 38
ETA-AAS: 53, 57, 60, 65
   Al in serum b:, 55
Excretion, of daily intake of Al: 13

Ferrioxidase, Al as inhibitor: 44
Ferritin: 43, 48
Fibrils, A4 amyloid: 23, 30
Fish, toxicity of Al to: 42
Flounder: 91
Flour: 57, 58, 61, 63, 64
   rice: 63
Fluid, cerebrospinal, Al in: 53
Fluids, protein infusion, contamination of: 52
Fluoride, effect on dissolution of Al: 72, 78
Fluorine: 16
Foil,
   containers: 90
   household: 90
Food,
   additives, Al in: 63, 73, 99
   Al contact with: 88
   contact with Al, Regulations: 92
   cooked in plain Al, Al levels of: 95
   determination of Al in: 52, 56
   significance of Al in: 63
Foodstuffs,
   Al levels in: 4
   migration of Al into: 95

Formulae, infant: 63, 70, 77, 81
Fruit: 69, 70
Fruit juice: 57, 58, 59, 63, 65

GFAAS; *see Graphite furnace atomic absorption spectrometry*
^{67}Ga: 28, 83
   as a marker: 27
Gastrointestinal tract: 65, 68, 83
   Al uptake via: 5
^{68}Ge, as probe for Si entry into rat brain: 28
Graphite furnace atomic absorption spectrometry: 28
Guam: 21
Gut,
   phosphate absorption from: 13
   sequestration of inorganic phosphate in: 14
   uptake of Al from: 74, 75

Haemodialysis: 52
Ham: 96
Hippocampus, Al content of: 24, 28
Huntington's disease: 20
Hyperphosphataemia: 13

ICP-MS; *see Inductively-coupled plasma mass spectrometry*
Imaging secondary ion mass spectrometry: 26, 31
Immunocytochemical staining: 29
Imogolite: 47
Inductively coupled plasma-mass spectrometry: 28, 83, 97
Infant formulae, Al in: 63, 70, 77, 81
Infants, bottle-fed: 68
Inhalation,
   Al intake via: 48
   of Al dust: 21
Inositol: 48
   phosphate: 45
Iron, in canned apple juice: 99

Ketones,
   as odour components: 98
   as taste components: 98

Maltol: 83

Margarine: 60, 62, 63, 70
Meat: 69, 70
Mechanism, of Al transfer to patient: 9
β-N-Methylamino-1-alanine: 21
Micronutrients, trace elements as: 17
Micropoisons, trace elements as: 17
Migration, of Al,
during food preparation: 99
during food storage: 99
Milk: 57, 59, 63, 97
Al in: 70, 74, 82
Ministry of Agriculture, Fisheries and Food: 5
Monitoring, of Al levels in blood: 10
Monomers, as coatings: 91
Motor neurone disease: 20

NAA; *see Neutron activation analysis*
Nasal-olfactory system, entry via: 48
Neurodegenerative disorders: 20
Neurofibrillary tangles: 21, 22, 23, 29, 37, 41, 77
Neutron activation analysis: 7
Newcastle bone disease: 7
Nickel: 16
Nutrition, parenteral: 52

Onion: 69
Oral intake of Al: 7
Osmosis, reverse: 40
of tap water: 8
Osteitis fibrosa: 7
Osteodystrophy, dialysis: 77
Osteomalacia: 7
Osteomalacic dialysis osteodystrophy: 40
Osteosclerosis: 7
Oxidation, anodic: 90
Oxide layer: 89
Oxides, of aluminium: 1

Packaging material: 88
Al as: 93
increased use of: 93
Parkinson's disease: 20
Pear: 70
Peptide, A4 amyloid: 29
Phosphate: 41
absorption: 28
absorption excluders: 1
binder: 77
binding of Al by: 44
groups: 42, 48
inorganic, sequestration of in gut: 14
intestinal absorption of: 53
Phosvitin: 45, 46
Plaques, senile: 22, 29, 37, 41, 46, 77
Plasma, Al concentration in: 2, 10
Pork: 70
'Positive lists', of coatings: 91
Potatoes: 57, 63, 69, 78, 97
mashed: 91
Powder, custard: 63, 64
Prediction,
of biological properties: 17
of toxicity: 17
Premature babies, bottle feeds for: 5
Protein,
A4 amyloid precursor: 30, 33
β-amyloid: 22, 23
Fe^{3+}-binding: 41
Psychomotor defects in the brain: 16

Rabbits, Al-induced tangles in: 24
Receptors, transferrin: 28
Regulations, food, contact with Al and alloys: 92
Renal dialysis: 2, 5, 77
Renal disease, chronic: 6
Renal failure, chronic: 28
complexity of: 16
Rhubarb: 72, 78
Rice: 69, 96
flour: 63

SIMS; *see Scanning electron microprobe X-ray analysis*
SIMS, imaging: 32
Saucepans, aluminium: 77

Scanning electron microprobe X-ray analysis: 25, 29
Sea water, Al in: 17
Selenium: 16
reduction in: 16
Senile dementia: 21, 77
Senile plaque core
Al deposits on: 26
Si deposits on: 26
Senile plaques: 22, 29, 37, 41, 46, 77
Septum: 28
Serum
Al in: 10, 54
Al in, by ETA-AAS: 55
Signals, atomic absorption: 58
Silicates, aluminium: 1
Silicic acid: 41, 42, 44, 47
polymerization of: 33
Silicon: 16, 30
as essential trace element: 40
deposits on senile plaque core: 26
uptake by the brain: 26
Sodium aluminium phosphate: 73, 99
Sodium aluminium silicate: 99
Soil, contamination by, 74
Solubility, of oxide on Al: 89
Soy meal, Al in: 82
Soya bean: 70
Spaghetti: 96
Spectroscopy, emission: 53
Staining, immunocytochemical: 29
Steel, stainless: 90
Studies, epidemiological: 33
Surface oxide layer: 88

Tangle-bearing neurones, calcium in: 21
Tangles, neurofibrillary: 21, 22, 23, 29, 37, 41, 77
Tap water: 6
reverse osmosis of: 8
Tea: 57, 58, 64, 69, 70, 77, 84
Al in: 5, 56, 77, 80
leaves: 57
Teflon: 90
Tin, in canned apple juice: 99
Tomatoes: 78
Toxicity,
of Al to fish: 42
of Al to trees: 42
of Al: 53
prediction of: 17
Trace elements: 5, 16
as micronutrients: 17
as micropoisons: 17
Tract, gastrointestinal: 65, 68, 83
Transferrin: 8, 26, 32, 41, 42, 43, 48
receptor, density of: 47
receptors: 28
Transport, of Al: 48
Trees, toxicity of Al to: 42
Turkey: 91

Urine: 56
excretion of Al in: 65
levels of Al in: 85

Vegetables, Al in: 69

Water,
drinking, Al from: 37
supplies, Al in: 32
tap: 6
tap, Al in: 97
Wheat: 69
Wine, white: 98
elemental content on storage: 98
World-wide consumption, of Al: 2

Zinc: 16